Geoscience Data and Collections

NATIONAL RESOURCES IN PERIL

Committee on the Preservation of Geoscience Data and Collections

Committee on Earth Resources

Board on Earth Sciences and Resources

Division on Earth and Life Studies

NATIONAL RESEARCH COUNCIL
OF THE NATIONAL ACADEMIES

THE NATIONAL ACADEMIES PRESS
Washington, D.C.
www.nap.edu

NOTICE: The project that is the subject of this report was approved by the Governing Board of the National Research Council, whose members are drawn from the councils of the National Academy of Sciences, the National Academy of Engineering, and the Institute of Medicine. The members of the committee responsible for the report were chosen for their special competences and with regard for appropriate balance.

This study was jointly sponsored by the American Association of Petroleum Geologists, American Association of Petroleum Geologists Foundation, American Geological Institute, Department of Energy–Fossil Energy (DE-AP75-00SW48036), Department of Energy–Yucca Mountain (DE-FG09-97NV12056), Geological Society of America, National Science Foundation (EAR-0071061), Paleontological Society (00-22-US-3548), Petrotechnical Open Software Corporation, Schlumberger, Ltd, Smithsonian Institution, and U.S. Geological Survey (00HQAG0145). Any opinions, findings, conclusions, or recommendations expressed in this publication are those of the authors and do not necessarily reflect the view of the organizations or agencies that provided support for this project.

International Standard Book Number: 0-309-08341-9

Library of Congress Control Number: 2002106850

Additional copies of this report are available from:

The National Academies Press
500 Fifth Street, N.W.
Box 285
Washington, DC 20055
800-624-6242
202-334-3313 (in the Washington metropolitan area)
http://www.nap.edu

Front cover: Geoscience data and collections examples and storage facilities. *Background left:* Flexible-space shelving at Bureau of Economic Geology, University of Texas at Austin. SOURCE: David Stephens, BEG, University of Texas at Austin. *Background right top:* Inside the National Ice Core Laboratory, at the Denver Federal Center in Lakewood, Colorado. SOURCE: Geoffrey Hargreaves, NICL. *Background right bottom:* Interior of the Ocean Drilling Program's Gulf Coast Repository (GCR) at Texas A&M University in College Station. SOURCE: Ocean Drilling Program. *Foreground left to right:* Fossil fish and trilobite. SOURCE: ExxonMobil Upstream Research Company; Foraminifera microfossils. SOURCE: ExxonMobil Upstream Research Company; Rock and mineral specimens. SOURCE: ExxonMobil Upstream Research Company; and Tapes containing data from boreholes. SOURCE: Phillipe Theys, Schlumberger, Ltd., Sugarland, Texas.

Backcover: *Background top:* Inside the National Ice Core Laboratory, at the Denver Federal Center in Lakewood, Colorado. SOURCE: Geoffrey Hargreaves, NICL. *Background bottom:* Interior of the Ocean Drilling Program's Gulf Coast Repository (GCR) at Texas A&M University in College Station. SOURCE: Ocean Drilling Program. *Foreground:* Fossil fish and trilobite. SOURCE: ExxonMobil Upstream Research Company.

Cover designed by Van Nguyen

Copyright 2002 by the National Academy of Sciences. All rights reserved.

Printed in the United States of America

THE NATIONAL ACADEMIES
Advisers to the Nation on Science, Engineering, and Medicine

The *National Academy of Sciences* is a private, nonprofit, self-perpetuating society of distinguished scholars engaged in scientific and engineering research, dedicated to the furtherance of science and technology and to their use for the general welfare. Upon the authority of the charter granted to it by the Congress in 1863, the Academy has a mandate that requires it to advise the federal government on scientific and technical matters. Dr. Bruce M. Alberts is president of the National Academy of Sciences.

The **National Academy of Engineering** was established in 1964, under the charter of the National Academy of Sciences, as a parallel organization of outstanding engineers. It is autonomous in its administration and in the selection of its members, sharing with the National Academy of Sciences the responsibility for advising the federal government. The National Academy of Engineering also sponsors engineering programs aimed at meeting national needs, encourages education and research, and recognizes the superior achievements of engineers. Dr. Wm. A. Wulf is president of the National Academy of Engineering.

The **Institute of Medicine** was established in 1970 by the National Academy of Sciences to secure the services of eminent members of appropriate professions in the examination of policy matters pertaining to the health of the public. The Institute acts under the responsibility given to the National Academy of Sciences by its congressional charter to be an adviser to the federal government and, upon its own initiative, to identify issues of medical care, research, and education. Dr. Harvey V. Fineberg is president of the Institute of Medicine.

The **National Research Council** was organized by the National Academy of Sciences in 1916 to associate the broad community of science and technology with the Academy's purposes of furthering knowledge and advising the federal government. Functioning in accordance with general policies determined by the Academy, the Council has become the principal operating agency of both the National Academy of Sciences and the National Academy of Engineering in providing services to the government, the public, and the scientific and engineering communities. The Council is administered jointly by both Academies and the Institute of Medicine. Dr. Bruce M. Alberts and Dr. Wm. A. Wulf are chair and vice chair, respectively, of the National Research Council.

www.national-academies.org

COMMITTEE ON THE PRESERVATION OF GEOSCIENCE DATA AND COLLECTIONS

CHRISTOPHER G. MAPLES, *Chair,* Indiana University, Bloomington
WARREN D. ALLMON, Paleontological Research Institution, Ithaca, New York
KEVIN T. BIDDLE, Exxon Mobil Corporation, Irving, Texas
DONALD D. CLARKE, Department of Oil Properties, City of Long Beach, California
BETH DRIVER, National Imagery and Mapping Agency, Reston, Virginia
THOMAS R. JANECEK, Florida State University, Tallahassee
LINDA R. MUSSER, Pennsylvania State University, University Park
ROBERT W. SCHAFER, Mineral Exploration and Business Development Consultant, Salt Lake City, Utah
ROBERT M. SNEIDER, Robert M. Sneider Exploration, Inc., Houston, Texas
JOHN C. STEINMETZ, Indiana Geological Survey, Bloomington
SALLY ZINKE (until 8/2001), Ultra Petroleum, Inc., Englewood, Colorado

NRC Staff

PAUL CUTLER, Program Officer
MONICA LIPSCOMB, Research Assistant
TERESIA WILMORE, Project Assistant

COMMITTEE ON EARTH RESOURCES

SUSAN M. LANDON, *Chair,* Thomasson Partner Associates, Denver, Colorado
JAMES C. COBB, Kentucky Geological Survey, Lexington
VICKI J. COWART, Colorado Geological Survey, Denver
MURRAY W. HITZMAN, Colorado School of Mines, Golden
JAMES M. McELFISH, JR., Environmental Law Institute, Washington, D.C.
JOHN MURPHY, University of Pittsburgh, Pennsylvania
DIANNE R. NIELSON, Utah Department of Environmental Quality, Salt Lake City
THOMAS J. O'NEIL, Cleveland-Cliffs, Inc., Cleveland, Ohio
DONALD PAUL, ChevronTexaco, San Francisco, California
RUSSELL STANDS-OVER-BULL, Arrow Creek Resources, Pryor, Montana
R. BRUCE TIPPIN, North Carolina State University, Asheville
MILTON H. WARD, Ward Resources, Inc., Tucson, Arizona
LAWRENCE P. WILDING, Texas A&M University, College Station
PHILLIP MICHAEL WRIGHT, Idaho National Engineering and Environmental
    Laboratory, Idaho Falls

NRC Staff

TAMARA L. DICKINSON, Senior Program Officer
KERI H. MOORE, Staff Officer
KAREN L. IMHOF, Senior Project Assistant

BOARD ON EARTH SCIENCES AND RESOURCES

RAYMOND JEANLOZ, *Chair,* University of California, Berkeley
JILL BANFIELD, University of California, Berkeley
STEVEN R. BOHLEN, Joint Oceanographic Institutions, Washington, D.C.
VICKI J. COWART, Colorado Geological Survey, Denver
DAVID L. DILCHER, University of Florida, Gainesville
ADAM M. DZIEWONSKI, Harvard University, Cambridge, Massachusetts
WILLIAM L. GRAF, University of South Carolina, Columbia
RHEA GRAHAM, New Mexico Interstate Stream Commission, Albuquerque
GEORGE M. HORNBERGER, University of Virginia, Charlottesville
DIANNE R. NIELSON, Utah Department of Environmental Quality, Salt Lake City
MARK SCHAEFER, NatureServe, Arlington, Virginia
BILLIE L. TURNER, II, Clark University, Worcester, Massachusetts
THOMAS J. WILBANKS, Oak Ridge National Laboratory, Tennessee

NRC Staff

ANTHONY R. DE SOUZA, Director
TAMARA L. DICKINSON, Senior Program Officer
DAVID A. FEARY, Senior Program Officer
ANNE M. LINN, Senior Program Officer
PAUL M. CUTLER, Program Officer
LISA M. VANDEMARK, Program Officer
KRISTEN L. KRAPF, Staff Officer
KERI H. MOORE, Staff Officer
YVONNE P. FORSBERGH, Research Assistant
MONICA R. LIPSCOMB, Research Assistant
EILEEN McTAGUE, Reseach Assistant
VERNA J. BOWEN, Administrative Associate
JENNIFER T. ESTEP, Administrative Associate
RADHIKA CHARI, Senior Project Assistant
KAREN L. IMHOF, Senior Project Assistant
SHANNON L. RUDDY, Senior Project Assistant
TERESIA K. WILMORE, Project Assistant
WINFIELD SWANSON, Editor

Reviewers

This report has been reviewed in draft form by individuals chosen for their diverse perspectives and technical expertise, in accordance with procedures approved by the NRC's Report Review Committee. The purpose of this independent review is to provide candid and critical comments that will assist the institution in making its published report as sound as possible and to ensure that the report meets institutional standards for objectivity, evidence, and responsiveness to the study charge. The review comments and draft manuscript remain confidential to protect the integrity of the deliberative process. We thank the following individuals for their participation in the review of this report:

Peter Crane
Royal Botanic Gardens
Kew, Richmond, Surrey, United Kingdom

Gordon Eaton
USGS *emeritus*
Coupeville, Washington

Stan Eschner
Trio Petroleum Inc.
Bakersfield, California

William L. Fisher
The University of Texas
Austin

Alexander H. Flax
Consultant
Potomac, Maryland

Scott Hector
Carneros Energy, Inc.
Bakersfield, California

Robert Laing
ChevronTexaco
Pleasanton, California

David Simpson
IRIS Consortium
Washington, DC

Milton A. Wiltse
Alaska Geological Survey
Fairbanks

Although the reviewers listed above have provided many constructive comments and suggestions, they were not asked to endorse the conclusions or recommendations nor did they see the final draft of the report before its release. The review of this report was overseen by Raymond A. Price, Queen's University, Kingston, Ontario, Canada. Appointed by the National Research Council, he was responsible for making certain that an independent examination of this report was carried out in accordance with institutional procedures and that all review comments were carefully considered. Responsibility for the final content of this report rests entirely with the authoring committee and the institution.

Preface

On September 20, 1999, the National Research Council (NRC) received a letter from Dr. Philip D. Vasquez, Deputy Assistant Secretary for Natural Gas and Petroleum Technology, conveying the request of the U.S. Department of Energy (DOE) that the NRC establish a committee to determine the options and develop a strategy for the preservation and management of subsurface geoscience data. Because of the broad concern on this matter across the geoscience community, a wide range of sponsors supported the activities of the committee. These sponsors were American Association of Petroleum Geologists, American Association of Petroleum Geologists Foundation, American Geological Institute, Department of Energy–Fossil Energy, Department of Energy–Yucca Mountain, Geological Society of America, National Science Foundation, Paleontological Society, Petrotechnical Open Software Corporation, Schlumberger, Ltd., Smithsonian Institution, and U.S. Geological Survey.

The committee operated under the aegis of the Committee on Earth Resources, a standing committee of the Board on Earth Sciences and Resources. It carried out its work through 4 meetings, 6 site visits by the full committee, 6 site visits by subsets of the committee, and distribution and analysis of a questionnaire. A total of 39 state geologic surveys and 17 other entities responded to the questionnaire. A list of oral and written contributions to the committee is provided in Appendix B. The full committee visited the following sites: the Smithsonian Institution in Washington, DC; the U.S. Geological Survey in Lakewood, Colorado; the Denver Earth Resources Library in Denver, Colorado; the National Geophysical Data Center in Boulder, Colorado; the Bureau of Economic Geology, University of Texas at Austin; and C&M Storage Inc. in Schulenberg, Texas. Subsets of the committee visited the Colorado School of Mines Geology Museum; DOE's Yucca Mountain project in Nevada; the Energy Information Administration in Washington, DC; the National Archives and Records Administration in College Park, Maryland; the Northern Rockies Geologic Data Center, in Billings Montana; and the U.S. Army Corps of Engineers in Washington, DC.

In responding to DOE's request to determine the options and develop a strategy for the preservation and management of geoscience data, the committee paid particular attention to the preservation and management of physical data (e.g., cores, cuttings, magnetic tapes, paper logs, rocks) as opposed to digital data. It is beyond the charge of the committee to focus on digital data. However, in keeping with the original intent of several funding agencies, the committee task was expanded beyond the original DOE request of "subsurface geoscience data" to include collections, especially those of a paleontological nature. It is important to clarify what is encompassed by the phrase "geoscience data and collections." "Geoscience" is a term for the collective subdisciplines of the geological (solid Earth) sciences, including geobiology, geochemistry, geohydrology, geophysics, sedimentology, and stratigraphy, among others. "Data" and "collections" were distinguished from each other on the basis of whether the physical item originated naturally (a rock, mineral, or fossil) or was produced from some other medium (a paper log, a magnetic tape, a picture); the former fell under the definition of collection and the latter fell under the definition of geoscience data (see Appendix D). The committee recognizes that the terms "collections" and "data" mean different things to different sectors of the geosciences. For example, the petroleum and mining industries consider rock cores and cuttings as "data," whereas the museum community considers them "collections." The definitions of these terms as used herein reflect the need for internal consistency within the report. In terms of geographic scope, the committee focused on geoscience data and collections of unconstrained geographic origin, but housed in the United States.

DOE's request to determine the options and develop a strategy for the preservation and management of geoscience data carries with it the implication that not everything can or should be preserved. To do otherwise is unrealistic and requires no determination of options—everything is kept. Consequently, the committee entered into this project with the assumption that not everything could or should be kept. However, the diversity and variety of geoscience data and collections are so vast that no specific set of protocols for obtaining or discarding geoscience data and collections applies in all cases. To that end, the committee has produced a set of guidelines under the premise that those who work with the appropriate geoscience data and collections (i.e., the user community) are the ones who are in the best position to assess which items to keep and which to discard.

The committee is indebted to the support and hard work of NRC staff. Teresia Wilmore (NRC Project Assistant) was very helpful in making sure the committee got to the right places and helped us with NRC travel and reimbursement. Monica Lipscomb (NRC Research Assistant) was instrumental in tracking down information and assisting with editorial copy after editorial copy. Paul Cutler (NRC Study Director) kept the committee on track, provided extremely useful summaries of complex discussions, reminded us of our tasks and obligations, and did the initial writing for many parts of the written document. Anthony de Souza (BESR Director) and Tamara Dickinson (NRC Senior Program Officer) provided very useful feedback and comments on rough drafts. Winfield Swanson (NRC Editorial Consultant) edited the first and last drafts.

Christopher G. Maples, Chair

Contents

EXECUTIVE SUMMARY   1

    The Need for Geoscience Data and Collections, 1
    Geoscience Data and Collections at Risk, 2
    The Charge to the Committee, 2
    What Should Be Preserved?, 2
    Options for Long-term Archiving and Access to Geoscience Data and Collections, 4
    The Time Is Now, 7

1 INTRODUCTION   8

    Intent and Outline of This Report, 8
    What Are Geoscience Data and Collections and Why Are They Important?, 9
    Investment in and Commercial Value of Geoscience Data and Collections, 14
    Users and Beneficiaries of Geoscience Data and Collections, 18

2 NATURE OF THE CHALLENGE   22

    Volume of Geoscience Data and Collections, 22
    A National Shortage of Space, 25
    Additional Sources of Loss of Geoscience Data and Collections, 27
    Inaccessible Geoscience Data and Collections, 32
    Priorities for Preservation of Geoscience Data and Collections, 34

3 GEOSCIENCE DATA AND COLLECTIONS TODAY   40

    Introduction, 40
    Cores and Cuttings, 40
    Media Containing Subsurface Data, 46
    Paleontological Collections, 50
    Rock and Mineral Collections, 53
    Other Data and Documentation, 54
    Summary, 56

4 MANAGING GEOSCIENCE DATA AND COLLECTIONS: CHALLENGES AND PRACTICES   57

    Introduction, 57

    Storage of Geoscience Data and Collections, 57
    Curation of Geoscience Data and Collections, 59
    Cataloging and Indexing, 63
    Access, 64
    Discovery and Outreach, 67
    Summary, 68

5 REGIONAL CENTERS: A MODEL FOR THE FUTURE   70

    Partnerships and Consortia, 70
    Repository Alternatives: Is One Too Few? Are 100 Too Many?, 71
    The Regional Centers Concept, 73
    Additional Roles and Responsibilities of the Federal Government, 77
    Incentives, 79

6 CHALLENGES AND SOLUTIONS   80

REFERENCES   82

APPENDIXES

A BIOGRAPHICAL SKETCHES OF COMMITTEE MEMBERS   87

B PRESENTATIONS TO THE COMMITTEE   90

C QUESTIONNAIRE   96

D TYPES OF GEOSCIENCE DATA AND COLLECTIONS   98

E GLOSSARY   99

F ACRONYMS AND ABBREVIATIONS   102

G NSF DIVISION OF EARTH SCIENCES (EAR) GUIDELINES FOR GEOSCIENCE DATA AND COLLECTIONS PRESERVATION AND DISTRIBUTION   104

H WEB SITES   106

Figures, Tables, and Sidebars

FIGURES

1-1   Examples of Geoscience Samples and Collections that Provide the Underpinnings for Geoscience Data, 10
1-2   Examples of Geoscience Data, 11
1-3   Hazard Zones for Lava Flows, Lahars, and Pyroclastic Flows from Mt. Rainier, Washington, 13
1-4   Flames from Hutchinson, Kansas, January 17, 2001, 20

2-1   1,000 feet (333 boxes) of Rock Core Laid Out Inside the Bureau of Economic Geology Core Facility, University of Texas at Austin, 24
2-2   Percentage of Available Space for Cores and Samples at State Geological Surveys, 25
2-3   Cost of Archiving Geoscience Data and Collections Versus Total Amount of Material Retained, 28
2-4   Staffing-Level Trends in the USGS's Geologic Division and Minerals Resource Survey Program (MRSP) from 1985 to 1996, 34

3-1   Examples of Where Geoscience Data and Collections Are Housed, 42
3-2   Coring and Cutting Devices, 43
3-3a Cores from Potter Mines, Matheson, Ontario, 43
3-3b Cuttings, 43

4-1   Ocean Drilling Program (ODP) Management Structure, 58
4-2   Onsite Study and Screening Space for Core at the Bureau of Economic Geology, University of Texas at Austin, 59

FIGURES FOR SIDEBARS

1-3   Collapsed Bridge Following Northridge Earthquake, 16
1-6   Fluid Inclusion Containing a Bubble of Gas, 20
1-7a Scientist Cleaning a Piece of Ice Core in a Cold Clean Room, 21
1-7b Close-up of an Ice Core from the Greenland Ice Sheet Project, 21

2-8   *Megalonyx jeffersoni* in the Indiana University Museum in the Late 1800s, 33
2-10 Cores Stored Outside at Oak Ridge National Laboratory, Tennessee, in Spring 2001, 36
2-11 Inside the National Ice Core Laboratory, 38

3-1   C&M Storage Inc. from the Air, 44
3-3   Interior of the Ocean Drilling Program Gulf Coast Repository in College Station, Texas, 47
3-4   Bureau of Economic Geology, University of Texas at Austin, 48
3-6   Building Stone Exposure and Test Wall, National Bureau of Standards, Washington, D.C., 53

TABLES

ES-1 Criteria for Determining Which Geoscience Data and Collections to Preserve, 3

1-1 Techniques for Recovering Oil Remaining after Primary and Waterflood Recovery, 18
1-2 Users and Beneficiaries of Geoscience Data, 19

2-1 Minimum Estimates of the Volume of Geoscience Data and Collections in the United States, 23
2-2 Examples of Transfer of Cores from Corporate-Owned Repositories to State Geological Surveys, 25
2-3a Available Space and Refusal of Samples at 35 State Geological Surveys, 26
2-3b Repository Space for Long-term Archiving of Geoscience Data and Collections at a Cross-section of Non-State Geological Facilities in the United States, 26
2-4 Threats to Geoscience Data and Collections, 28
2-5 Criteria for Determining Which Geoscience Data and Collections to Preserve, 37
2-6 Guidelines for Assessing Donation and Reception Priorities for Donors and Recipients of Geoscience Data and Collections, 39

3-1 Examples of Collectors of Geoscience Data and Collections, and Their Purpose, 41
3-2 Physical Parameters Recorded in Well Logs, 50
3-3 The 17 Largest Fossil Collections in the United States, 51
3-4 Paleontological Collections in the United States at Risk of Becoming Endangered or Orphaned in the Next Decade, 51
3-5 Holdings of the National Mine Map Repository, 55

4-1 Libraries and Geologic Sample Repositories—A Comparison of Cataloging Practices, 63
4-2 Incentives for Improving the Ability to Find Information about Geoscience Data and Collections, 67

5-1 Qualitative Assessment of Repository Options, 71
5-2 Percentage of Total U.S. Oil Production, 1945–1975 and 1976–2000, as a Proxy for Volume of Geoscience Data and Collections in the Gulf Coast, Pacific Coast, and Rocky Mountain Regions, 75
5-3 Estimated Cost Range to Establish a Regional Center, 76
5-4 Estimated Range of Recurring Costs for Each of the Three Research Centers, 77
5-5 Representative Service Charges, 77
5-6 Proposed Roles of a Federal Geoscience Data and Collections Coordinating Committee and Federal External Science-Advisory Boards, 78

D-1 Examples of Geoscience Collections, 98
D-2 Examples of Derived and Indirect Geoscience Data, 98

SIDEBARS

1-1 Statement of Task, 9
1-2 Increased Use of Core for Research Articles in the *Bulletin of the American Association of Petroleum Geology*, 14
1-3 Los Angeles Basin, 16
1-4 Extraction of Additional (By-passed) Oil and Gas Reserves, 17
1-5 Elmworth Gas Field, 19
1-6 A New Use of Cuttings: Analysis of Fluid Inclusions, 20
1-7 Ice Core Reuse, 21

2-1 Findings of the American Geological Institute in 1997, 24
2-2 Shell Oil's Donation of Geoscience Data: A Success Story in Texas, 27
2-3 Regrettable Losses, 29
2-4 Dibblee Foundation: Ensuring Knowledge Transfer, 29
2-5 Australian and Canadian Assessment Reporting Requirements: A Contrast to Those in the United States, 30
2-6 How Much Core and Cuttings Does the Average Minerals Exploration Project Produce?, 31

FIGURES, TABLES, AND SIDEBARS

2-7 Extracts from an E-mail Notice Sent by Killam Associates of Millburn, New Jersey, to the Paleontological Research Institution, Ithaca, New York (October 25, 2001), 32
2-8 Examples of "Lost" Fossils, 33
2-9 The National Museum of Natural History and the U.S. Geological Survey, 35
2-10 Examples of Inaccessible Geoscience Data and Collections, 36
2-11 Managing Ice Cores at the National Ice Core Laboratory, 38

3-1 C&M Storage Inc., 44
3-2 USGS Core Research Center at the Denver Federal Center, Lakewood, Colorado, 45
3-3 Ocean Drilling Program Facilities, 47
3-4 Bureau of Economic Geology, University of Texas at Austin, 48
3-5 Alaska Geologic Materials Center, 49
3-6 The Merrill Collection of Building Stones, 53
3-7 Reno Sales–Charles Meyer–Anaconda Memorial Collection, 54
3-8 Examples of Government Holdings of Documentation, 55
3-9 Denver Earth Resources Library, 56

4-1 Calgary Core Research Centre, 60
4-2 National Geophysical Data Center, Marine Geology and Geophysics Division, 62
4-3 Profiling the Collections at the Smithsonian: A Tiered Approach to Collections Description, 64
4-4 Institute of Museum and Library Services, 65
4-5 Wyoming Oil and Gas Conservation Commission, 66
4-6 Method of Attribution for Reports Using Ocean Drilling Program Data and Collections, 68
4-7 Geoinformatics, 68
4-8 Smithsonian's Research and Collections Information System, 69

Executive Summary

Everyone in downtown Hutchinson, a city of 40,000 in central Kansas, heard or felt the explosion, Wednesday morning, January 17, 2001. Natural gas burst from the ground under Woody's Appliance Store and the adjacent Décor Shop, blowing out windows in nearby buildings. Within minutes, the two businesses were ablaze. That evening, geyser-like fountains of natural gas and brine, some reaching heights of 30 feet, began bubbling up 3 miles east of the downtown fires. The next day, natural gas, migrating up a long-forgotten brine well, exploded under a mobile home and killed two people. The city ordered hundreds of residents to evacuate homes and businesses, many of whom would not be able to return until the end of March (Allison, 2001[1]).

The Kansas Geological Survey (KGS) stepped into a situation where demand for answers was great, but information was in short supply. Fortunately, the KGS had cores preserved in its repository from a project the Atomic Energy Commission had conducted in the 1960s to investigate the geology of localities being considered for nuclear storage. Practically unused for more than 30 years, these cores contained information that could be obtained rapidly—and without the time or risk of drilling into another unknown gas pocket. Geologists examined these and other cores and samples from wells drilled in the area to get a sense of the potential paths for gas flow through the rock. Armed with this information, obtained using geoscience data and collections, the KGS gathered new seismic data around the city, from which two anomalous zones of potential high gas pressure were identified. The gas had migrated 8 miles from a leaking salt cavern used as an underground natural gas storage facility. This gas was then safely vented. Over the next two months the Kansas Gas Service consulted with the KGS about possible vent-well locations and additional vent wells were drilled to release pressure. Hutchinson was safe from further gas geysers and gas explosions—and the displaced residents finally could return safely to their homes. Understanding of the situation was initiated through the KGS's fast action—action that began with *cores* that had been collected for another purpose many years earlier. Having immediate access to critical geoscience data and information played a crucial role in facilitating rapid response to a local crisis.

THE NEED FOR GEOSCIENCE DATA AND COLLECTIONS

This report builds the reader's understanding of the utility of geoscience data and collections, why these were acquired initially, why many remain useful, and what should be kept. ***Geoscience data and collections***[2] (e.g., ***cores, cuttings, fossils,*** geophysical tapes, paper logs, rocks) are the foundation of basic and applied geoscience research and education, and underpin industry programs to discover and develop domestic natural resources to fulfill the nation's energy and mineral requirements. Geoscience data and collections record the history of processes that operate on the Earth today and in the past and provide insights that lead to improved prediction of hazards, both immediate and long term. The geoscience community has amassed an enormous wealth of data and collections, most of which remain potentially useful and would be costly to replace, and much of which cannot be replaced. The diversity and quantity of these geoscience data and collections continue to expand, and as

---

[1]Reprinted with permission from *Geotimes*, October 2001. Copyright American Geological Institute, 2001.

[2]Geoscience collections are groupings of individual geoscience items that may be related by sample type, geographic location, or scientific or applied interests (see Appendix E for more information on this and other technical terms **highlighted** in the text).

they have, so has need for space and funding to support their *preservation* and accessibility.

Archiving and maintaining data and materials collected during the course of geoscientific research carry benefits well beyond those recognized by the scientific and academic communities. Well-maintained and well-documented geoscience data and collections are storehouses of information that likely will result in better assessment and management of natural resources, better understanding of the geologic hazards with which we live, and enhanced knowledge of the history of Earth and life. Virtually every facet of our daily life is touched either directly or indirectly by geoscience data and collections—from power that lights our cities to coatings on paper in books to medicines that save lives. If you drive a car, ride a bus, walk on sidewalks, take medicine, wear synthetic fabrics, or read a magazine, you have come in direct contact with and used geoscience resources, all of which owe their origin to information gleaned from geoscience data and collections.

Both the quality and quantity of geoscience data and collections have direct bearing on the accuracy of predicting and meeting future resource and engineering needs. Moreover, geoscience data and collections provide critical information that scientists and engineers need to help inform a variety of important societal decisions, including problems resulting from increased population growth on our planet. For example, current fossil energy resource assessment and exploitation is based directly on knowledge of the subsurface geological and engineering properties of the rocks that contain the resources. Natural hazards are assessed using historical records of their occurrence, coupled with prehistoric evidence gathered using geoscience data and collections. In both cases, absence of geoscience data and collections means that interpretations will be weaker at best and erroneous at worst.

GEOSCIENCE DATA AND COLLECTIONS AT RISK

Geoscience data and collections are imperiled, even though many are potentially useful and valuable in the future. Billions of dollars have been spent to acquire them. For instance, the U.S. Geological Survey (USGS) estimates that the cost to replace the geoscience data and collections archived in its Core Research Center at Lakewood, Colorado—a facility that contains no more than 5 percent of the volume of at-risk geoscience data and collections in the United States—is on the order of $10 billion (NRC, 1999a). Other examples include federal support in excess of $500 million for the acquisition of deep-sea sediment cores by the Ocean Drilling Program between 1983 and 1998 (NRC, 2000), and the estimated $535 million value of geologic materials housed at the Kentucky Geological Survey (Kentucky Geological Survey, 2001).

The committee learned that many geoscience data and collections already have been lost, and many more are at risk. Housing of and access to geoscience data and collections have become critical issues for federal and state agencies, academic institutions, museums, and industry. Nearly two-thirds of the state geological surveys the committee polled reported that their geoscience data and collections libraries have 10 percent or less space remaining for new data and collections. Even more critical, 46 percent of those same state geological surveys either reported that there is no space available or have refused to accept new material.

THE CHARGE TO THE COMMITTEE

The dilemma over geoscience data and collections is this: more and better geoscience data and collections exist now than ever before, however planning for space and maintenance of these data and collections have not kept pace with their acquisition. Therefore, appropriate management of these data and collections has become a more critical problem now than ever before. Consequently, the overall goal of this study was to develop a comprehensive strategy to manage geoscience data and collections in the United States. Specifically, the committee was charged with the following tasks:

1) Develop a strategy for determining which geoscience, *paleontological*, *petrophysical*, and engineering data to preserve.
2) Examine options for the long-term archiving of and provision of access to these data.
3) Examine three to five *accession* and *repository* case studies as examples of successes and failures.
4) Distinguish the roles of public and private sectors in data preservation.

The committee concentrated its effort on the preservation and management of physical data (e.g., cores, cuttings, fossils, geophysical tapes, paper logs, rocks) as opposed to digital data (e.g., computer-stored information). Nevertheless, the committee addressed the use and importance of digital information to enhance cataloging and dissemination of information about the physical materials (i.e., metadata about the geoscience data and collections). Digital access to information about geoscience data and collections is a key ingredient to their use by the widest range of clients possible.

WHAT SHOULD BE PRESERVED?

Geoscience data and collections are valuable national resources, some of which should be preserved and made available for scientific and strategic use. Despite their importance, utility, and value, substantial amounts already have been lost. For example, the record of the deepest *well* cored in the United States has been lost. The present-day cost to acquire a similar core is estimated at $12.3 million to $16.4 million (Michael Padgett, EEX Corporation, personal communication, 2001).

TABLE ES-1 Criteria for Determining Which Geoscience Data and Collections to Preserve

Criteria	Well Documented[d]	Irreplaceable[e]	Potential Applications[f]	Accurate	Quality/ Completeness	Non-Replicative
Collections:						
Cuttings	X	x	x	X	—	X
Engineering[a]	X	x	x	X	x	—
Fossils	X	x	x	X	x	—
Geophysical[b]	X	x	x	—	x	X
Maps/Notes[c]	X	x	x	—	x	X
Mining Cores	X	x	x	X	x	—
Other Rock Cores	X	x	x	X	x	X
Sediment & Ice Cores	X	x	x	X	X	—

X=present or necessary for preservation (i.e., absence = candidate for deaccession).
x=may be present and may be a factor for preservation (i.e., absence may not be a factor for deaccession).
_=not present and not necessary for preservation (i.e., absence is not a factor in deaccession).
Criteria are arranged from left to right in approximately decreasing order of importance (but see text for further explanation and elaboration).
Collections are arranged alphabetically.

[a]Includes drill stem tests, completion records, site reports, and other engineering data/reports on CD, computer disk, fiche, paper, tape, or some other quasi-stable medium.

[b]Includes seismic data, down-hole geophysical data, fly-over geophysical data, and other geophysical data on CD, computer disk, fiche, paper, tape, or some other quasi-stable medium.

[c]Includes unpublished materials on CD, computer disk, fiche, paper, tape, or some other quasi-stable medium, whether or not they were used in the production of published products.

[d]All collections must be well documented before any other assessment of their utility and future can be done. Indeed, whether or not a rock, fossil, core, or other item is replaceable is completely unknown in the absence of adequate documentation to assess uniqueness. That said, if part of a collection is not replaceable, but only documented well enough to know that it is unique, it probably should be kept anyway. Documentation includes, but is not limited to, information about age, location, depth, collector or author, date acquired, and associated materials.

[e]Impossible or highly unlikely to collect a similar sample (e.g., a mine core from a completely mined-out locality; a sample from a politically inaccessible part of the world; a sample requiring great time and effort to recollect such as a deep ice core from Antarctica or Greenland).

[f]This category in particular should be weighed judiciously by a science advisory board comprised of members of the user community.

Potential causes of loss of existing materials are numerous. Examples include lack of space in repositories, changing interests of some companies away from domestic production, company mergers, deterioration of materials and accompanying information over time, changes in staff and staff research interests, and reductions in work force at government facilities. Based on information presented to us and gathered over the course of the study, *the committee concludes that many geoscience data and collections are currently in peril.* Therefore, **the committee recommends that priority for rescuing geoscience data and collections be placed on those that are in danger of being lost. The committee recommends that the highest priority for retention and preservation be directed toward data and collections that are well documented and impossible or extremely difficult to replace.** Other factors to consider when setting priorities for preservation are potential applications, accuracy, quality and completeness, and redundancy. Table ES-1 summarizes the committee's assessment of overarching factors pertinent to the decision to retain or discard (***deaccess***) geoscience data and collections.

Assessing potential applications of geoscience data and collections is an important step in prioritization, and is a challenge that should not be left to a single individual. Assessing basic and applied potential of any physical data is a task that requires vision, imagination, and broad experience. Such guidance should be sought through external science advisory boards that represent a broad range of scientific, government, and business communities (collectively, the user community). Examples of the user community advising on priorities for preservation include those for the National Ice Core Laboratory and the Ocean Drilling Program. Such advisory committees are in a position to provide realistic recommendations (as opposed to the unrealistic recommendation of "keep everything") about what to keep using criteria suggested above against a backdrop of what might be needed in the future.

Enormous volumes of geoscience data and collections are held by a large number and variety of institutions. Museums, state geological surveys, universities and colleges, federal agencies, and industry all hold geoscience data and collections that have been amassed over as many as several hundred years. The committee estimates that more than 15,000 miles of cores and cuttings, well over a quarter of a billion line-miles of seismic data, and more than 100 million boxes of fossils are in geoscience repositories today. Furthermore, *the committee concludes that sufficient geoscience data and collections in the United States are at risk of loss to fill at least 20 times the*

USGS Core Research Center in Lakewood, Colorado. These figures are estimates that reflect minimum values.

Assessing the complete breadth and depth of geoscience data and collections that exist was just one of the challenges the committee faced. Simply stated, the quantity, variety, and quality of the nation's geoscience data and collections are largely unknown. The committee found that information on geoscience data and collections that have been lost or discarded is elusive because of their proprietary nature, the unwillingness to admit to discarding such data and collections, and the challenges and costs of donating them to a public facility (for example, the ongoing 6-year-old negotiations between Shell Oil Company and the Internal Revenue Service) versus discarding them.

Consequently, the committee became keenly aware that an understanding of the wealth of geoscience data and collections available to the public and private science and technology sectors is imperative. Without that understanding, we cannot make the best use of what already exists, or understand what is now at risk of being lost or discarded. Gathering comprehensive information on existing data and collections is essential for their future use. Therefore, **the committee recommends funding cataloging efforts to gather comprehensive information about existing geoscience data and collections. The committee recommends that access to these funds be on a competitive basis, and that preference be given to institutions with holdings that meet the same priorities as those outlined above for preservation.** The Institute of Museum and Library Services and the National Science Foundation are two federal agencies with experience and demonstrated effectiveness at distribution of funds to the museum, library, and science communities on a competitive basis. The inventory process should proceed simultaneously with development of a geoscience data and collections management system, and, to stimulate knowledge and use of the data and collections, the resultant institutional catalogs should be available online.

The number of universities, colleges, museums, institutes, state agencies, and other geoscience-oriented entities that need support for these cataloging efforts is certainly in the hundreds. Therefore, **the committee recommends that this initial catalog funding effort target 5 to 10 institutions each year until the nation's geoscience data and collections are adequately assessed.** The committee estimates that this effort would be effective if supported at the level of $5 million to $10 million per year.

OPTIONS FOR LONG-TERM ARCHIVING AND ACCESS TO GEOSCIENCE DATA AND COLLECTIONS

Managing Geoscience Data and Collections in the United States

Because the volume and variety of geoscience data and collections are great, the goal of achieving long-term archiving of and access to geoscience data and collections must be achieved sequentially. **The committee recommends the establishment of a distributed network of regional geoscience data and collections centers, each with an external science advisory board.** Each center would be a consortium of government, academic, and industry entities within the region, and would likely build off existing infrastructure and expertise. Among their various roles, the centers would foster cooperation among existing repositories, encourage adoption of uniform standards, and coordinate outreach. The committee found that successful (i.e., supported, maintained, and used) geoscience data and collection centers served relatively focused communities of interest (most often geographically defined areas). An excellent example of such centers, with external science advisory boards, broad community involvement, and regional distribution can be found in the current core repositories for the Ocean Drilling Program.

There was consistency among those testifying to the committee, and consensus within the committee itself, that one model of a single, national geoscience repository was impractical. Four barriers stand out to such a model: the untenable cost of moving all geoscience data and collections to a single location, the enormity of scale that such a center would entail, the impracticalities of expecting many users to come to the center, and the unwillingness of many existing repositories to part with their collections. Regional centers, on the other hand, are large enough to achieve economies of scale, but small enough to encourage local interest and support. Distributing the centers would permit sponsors to nurture regional networks of dedicated volunteers, content donors, and financial benefactors.

The committee concludes that immediate action is needed to stop the loss of irreplaceable geoscience data and collections in areas containing the greatest volume of at-risk material. Criteria for assessing risk include those outlined earlier. In terms of sheer volume of data, shifting priorities of those holding data, and merger activity, those regions with long histories of resource extraction stand out. **The committee recommends establishing three centers (one each in the Gulf Coast, Rocky Mountain, and Pacific Coast regions). Furthermore, the committee recommends that additional regional centers, as merited, be established over the next 5 to 10 years, and that preference be given to centers that meet three main criteria: 1) need for such a center in the region (i.e., active clientele, identified collections of high priority, at-risk data in the region), 2) broad involvement and support among various regional geoscience and other entities (government, academia, and industry), and 3) active participation of an independent, external science-advisory board. The committee recommends that the centers build upon existing expertise and infrastructure—such as state geological surveys, museums, universities, and private enterprises—and that, where practical, more efficient use of existing space be** encouraged before expansion. Furthermore, **the committee recommends that access to the center-establishment and improvement funds be on a competitive basis.**

For reasons stated above, the National Science Foundation is a logical distributor of the funds. The committee estimates that each center would cost between $35 million and $50 million to establish.[3] Additional support would be needed for operations costs. Therefore, **the committee recommends additional maintenance and operations expenditures, which would be re-evaluated regularly on a competitive basis, to ensure maximum utilization of each center (i.e., to encourage public outreach and awareness, use, and cost-sharing activities).** The committee estimates these costs to be in the range of $3 million to $5 million per year for each center.

A Strategy for Managing Federal Geoscience Data and Collections

Federal agencies responsible for geoscience data and collections in the United States should lead the way by setting examples of good practices in preservation and use of geoscience data and collections. Such examples serve to promote public good, increase the visibility of the federal side in a leadership role, and increase the likelihood of federal partnerships with the private sector.

The committee learned that inadequate levels of support for cataloging and archiving of geoscience data and collections exist within many federal entities. For example, at the National Museum of Natural History (NMNH), which houses the nation's largest publicly available geoscience collection, only 10 percent of the holdings currently are electronically cataloged. Moreover, both the NMNH and the USGS have experienced reductions in force that have compromised their ability to care for their collections. Therefore, **the committee recommends that federal agencies be supported to the same extent as non-federal institutes and consortia with respect to cataloging and repositories, and with regular review. The committee recommends that priorities for federal agency support should closely follow those recommended for the regional centers: 1) need for such a repository in the agency, 2) broad or active involvement within and among various federal geoscience agencies (e.g., BLM, DOE, EPA, NASA, NOAA, NSF, USACE, USGS, USNM), and 3) active participation of independent, external science-advisory boards.** The committee envisages (where appropriate) federal agencies as potential members of the proposed regional consortia, with funding for federal and non-federal entities in this instance converging within these consortia. Such arrangements between state and federal agencies are already in place in Colorado and Alaska, for example. Lastly, federal agencies should be permitted to offset some costs with appropriate charges for selected services.

While it exists, coordination among federal agencies that collect or archive geoscience data and collections could be improved. Such improved coordination would optimize sharing of business practices and consumer use of related data collected by various agencies or establishing priorities among agencies so that limited funds can be used to the best overall effect. Adoption of consistent and good practices, along with a clarification of roles, would, at a minimum, increase efficiencies for federal agencies and the user community, comparable in some respects to the goals of the National Spatial Data Infrastructure (NRC, 1993) and the Geospatial One-Stop initiatives.[4] In addition, such collaboration would render the whole of government holdings more complete, enhance the value of individual components, and permit a significantly (and, eventually, measurable) increased benefit to diverse communities.

To optimize federal coordination, **the committee recommends establishing a federal geoscience data and collections coordinating committee.** Such a committee could be established and funded through the Office of Management and Budget, as the committee would oversee coordination and increased efficiency among a range of federal agencies. This federal geoscience data and collections coordination committee should be broad-based, reaching between and within all federal and quasi-federal agencies involved in geoscience research or geoscience data and collections acquisition. The committee's charge should focus on coordination of federal agencies' roles with regard to geoscience data and collections preservation, access, and use. **The committee recommends that the federal geoscience data and collections coordinating committee should appoint several federal external science advisory boards to advise on priorities for federal holdings, with respect to preservation, cataloging, and access among and within federal and quasi-federal agencies.** Previous NRC reports (e.g., NRC, 2001) already have noted the value for federal agencies of having direct external community involvement and advice to help set internal priorities for funding, monitoring, and research efforts. Examples of federal external science advisory boards that deal with collections are those within the operating structure of the National Ice Core Laboratory (coordinated jointly by the USGS and NSF) and the Smithsonian Institution.

The federal, external science advisory boards would focus on holdings within the federal government, but would

---

[3]The committee bases its estimates on building anew, and recognizes that costs could be less if a center were to build off existing infrastructure.

[4]These two initiatives are useful models in several respects. First, they seek to render data from many federal, state, and local agencies both convenient to access and easy to use together. Second, they must address diverse missions, user communities, producer concerns, data definitions, and data formats. Information providers may themselves produce the data, or they may obtain data from external sources. Coordination of U.S. geoscience data and collections will involve all of these issues.

coordinate with the science advisory boards recommended for the regional geoscience data and collection centers. The federal, external science advisory boards, which could be discipline-based, would advise on establishment of consistent practices across agencies with respect to preservation of and access to geoscience data and collections acquired from public lands or using federal funds. In addition, the federal, external science advisory boards would advise on what geoscience data and collections should logically fall within the purview of various federal agencies. Monitoring of conformance to agreed-upon practices, as a question of how rather than what, would reside within the charge of the federal geoscience data and collections coordinating committee.

The federal geoscience data and collections coordinating committee would have other responsibilities related to how the federal effort should be streamlined, coordinated, and improved. One such responsibility would be monitoring implementation of electronic reporting for all exploration, exploitation, and research reports currently submitted to the federal government. The committee believed that electronic reporting was a necessary step to minimize the burden of cataloging newly collected geologic data and samples, while maximizing their potential use. As noted, the challenge to catalog existing geoscience data and collections is already immense. Therefore, **the committee recommends that electronic reporting be implemented as soon as possible, with additional funding as required to accelerate it.** Examples of programs of electronic reporting can already be found at the provincial level in Canada and Australia, and in the state of Wyoming.

The cataloging effort recommended for non-federal institutional holdings is of equal importance for future use of federal geoscience data and collections. Therefore, **the committee recommends that the federal geoscience data and collections coordinating committee monitor and facilitate progress of cataloging efforts across the federal government.** Here, the federal geoscience data and collections coordinating committee should work closely with the federal, external science advisory boards to determine which cataloging efforts warrant the highest priority. In addition, the federal geoscience data and collections coordinating committee should facilitate and coordinate Internet access to all federal geoscience data. This would include (but not be limited to) reports and catalogs of holdings, location and availability of similar geoscience data and collections, and contact information (where appropriate) for onsite use of geoscience data and collections. Success of this effort will be enhanced by coordinated adoption of digital data standards to improve interoperability of interagency information.

Regular review of the roles of the National Science Foundation and Institute of Museum and Library Services as distributors of funds for non-federal cataloging and repository efforts is essential. If existing external review mechanisms (e.g., committees of visitors, external steering committees) are inadequate for this task, new ones should be devised.

The Roles of Public and Private Sectors

From the testimony of those who use geoscience data and collections (see Appendix B) *the committee concluded that incentives (and even some mandates) for preservation of geoscience data and collections would encourage preservation efforts, and that partnerships and consortia are the most appropriate means by which to maintain long-term security for the various regional repositories.* Therefore, **the committee recommends establishing a combination of federal, state, regional, and local government incentives and requirements for geoscience data and collections donations and deposition. Establishing such incentives should be an immediate priority to stem the tide of lost and discarded geoscience data and collections, many of which remain useful.** Such incentives would encourage private donations of geoscience data and collections by providing credit for shipping costs and fundamental recognition that fossils, rock, sediment, and ice are unique and have donation value. When such data and collections are used to enhance recovery of resources, federal support for these incentives could pay for itself many times over (see DOE, 2002). An incentive for the research community would be a requirement that geoscience data and collections amassed during federally funded research (i.e., funded by agencies such as DOD, DOE, EPA, NASA, NSF, USGS, USNRC) be archived appropriately, cataloged, and made accessible to the public (e.g., NSF guidelines in Appendix G, and in USGCRP, 1991). Federal support for research should be, in general, contingent upon the public availability of these geoscience data and collections within a reasonable time.

The geoscience community itself must take more responsibility for preservation and use of geoscience data and collections. Although the necessity and importance of these data for research and interpretations are broadly accepted, adequate curation and long-term care for them take time and consequently fall through the cracks. The geoscience community should do more than just acknowledge the importance of geoscience data and collections—it should establish incentives, rewards, and requirements for their care and accessibility. **The committee recommends that the geoscience community adopt standards for citation in scientific and other publications of geoscience data and collections used.** Citation histories enhance credibility and importance to well-organized, often-used data and collections. In addition, **the committee recommends that institutions and professional societies establish (where appropriate) awards and other forms of recognition for outstanding contributors to the preservation and accessibility of geoscience data and collections.**

EXECUTIVE SUMMARY

THE TIME IS NOW

Well-maintained and well-documented geoscience data and collections have both immediate and long-term value. The nation has assembled a wealth of geoscience data and collections. Some of these already have been lost, and many more are in imminent danger of being lost—through mismanagement, neglect, or outright disposal—if immediate action is not taken. The recommended solutions that this committee proposes represent a strategy for such immediate action. Future generations deserve the opportunity to build upon existing successes and avoid repetition of our failures. Geoscience data and collections are national resources, and are a part of our nation's heritage. Preservation of geoscience data and collections is a comparatively small investment in our past, our present, and our future, with both immediate and long-term benefits.

1

Introduction

INTENT AND OUTLINE OF THIS REPORT

Geoscience data and collections[1] (e.g., ***cores, cuttings, fossils,*** geophysical tapes, paper logs, rocks) are the foundation of basic and applied geoscience research and education, and underpin ***industry*** programs to discover and develop domestic natural resources to fulfill the nation's energy and mineral requirements. Geoscience data and collections record the history of processes that operate on the Earth today and in the past and provide insights that lead to improved prediction of potential hazards, both immediate and long term. The geoscience community has amassed an enormous wealth of data and collections, most of which remains potentially useful and would be costly to replace, and much of which is irreplaceable. The diversity and quantity of these geoscience data and collections continue to expand, and as they have, so has need for space to support the preservation of and access to those needing preservation.

Because the ability to preserve and maintain geoscience data and collections has not kept pace with the generation and acquisition of new data through the decades (as demonstrated within this report through examples of data loss and lack of space), the nation is now in danger of irretrievably losing valuable and unique geologic records. The United States has not planned for the large amount of geoscience data that merits preservation. Yet geoscience data and collections are a trove of untapped resources awaiting scientists, engineers, educators, and policy makers who can consolidate and use the information. It is an appreciation of these issues that led a broad range of government agencies and other organizations (American Association of Petroleum Geologists, American Association of Petroleum Geologists Foundation, American Geological Institute, Department of Energy–Fossil Energy, Department of Energy–Yucca Mountain, Geological Society of America, National Science Foundation, Paleontological Society, Petrotechnical Open Software Corporation, Schlumberger, Ltd., Smithsonian Institution, and U.S. Geological Survey) to support this study. The statement of task to the study committee is given in Sidebar 1-1.

This report builds the reader's understanding of the utility of: geoscience data and collections; why the data and collections were acquired initially, why many remain useful, and what should be kept (item 1 of the committee's charge); the magnitude of physical (as opposed to digital) geoscience data and collections that exist and where they currently reside (item 2 of the committee's charge); and the difference between space available and space needed to retain and properly use many of these geoscience data and collections. These topics are covered in the first three chapters of the report. Chapter 4 demonstrates the necessary steps in preserving geoscience data and collections (item 3 of the committee's charge), and Chapter 5 examines potential roles of the public and private sectors in preservation (item 4 of the committee's charge), and maps out a national strategy for effectively managing geoscience data and collections (the overall goal of the committee's charge).

From the outset, it is important to understand what is meant by the term preservation. Preservation involves a number of interrelated processes: evaluation of what should be kept, acquired, or assimilated; organization and maintenance, both of the physical samples, and of supporting information, or ***metadata***; making users aware of geoscience data and collections; making geoscience data and collections acces-

---

[1]Technical terms are defined in Appendix E and **highlighted** upon first occurrence in the report. The committee defines **geoscience** as the collective disciplines of the geological sciences, including engineering geology, geobiology, geochemistry, geohydrology, geophysics, sedimentology, and stratigraphy, among other solid-Earth-based subdisciplines. This definition contrasts with Earth science, which the committee defines as the collective disciplines of whole-Earth study, including atmospheric science, ocean science, and geoscience. Geoscience collections are groupings of individual geoscience items that may be related by sample type, geographic location, or scientific or applied interests.

> **SIDEBAR 1-1**
> **Statement of Task**
>
> The preservation of geoscience data (e.g., cores, cuttings, maps, paper reports, digital data[a]) is becoming a critical issue for federal agencies, academic researchers, museums, institutes, and industry. This study will:
>
> (1) develop a strategy for determining what geoscience, paleontological, petrophysical, and engineering data[b] to preserve;
> (2) examine options for long-term archival of these data;
> (3) examine three to five accession and repository case studies as examples of successes and failures; and
> (4) distinguish the roles of the public and private sectors in data preservation.
>
> The overall goal of the study is to develop a comprehensive strategy for managing geoscience data in the United States.
>
> ---
>
> [a]The committee chose to emphasize physical data (as opposed to digital data) in its considerations because the preservation of physical data presents more of a challenge within the geosciences, and numerous digital data initiatives and studies either have been completed recently or currently are underway.
> [b]For the sake of clarity and simplicity, the committee, using the definition of geoscience noted earlier, considered paleontological, petrophysical, and engineering data that related to solid-Earth studies to be part of geoscience data and collections. Geoscience data and collections were distinguished from each other on the basis of whether or not the physical item or items originated naturally (a rock, mineral, or fossil) or were produced from some other medium (a paper log, a magnetic tape, a picture), with the former falling under the definition of collection and the latter falling under the definition of geoscience data. See Appendix D for examples and clarification, and Appendix E for a glossary of technical terms.

sible to users; and making samples and data useful and of sufficient quality and validity to be believable. A successful strategy for managing geoscience data and collections in the United States must address all components. First and foremost, however, these components rest on a single, critical element—good accompanying documentation[2] for the data and collections.

Proper *curation* of geoscience data and collections is more efficient and less redundant than repeatedly re-collecting the samples. Archiving costs[3] summed over many decades may approach reacquisition costs, but the value of ready access to data and collections (for hazards response and other unanticipated uses, education, and academic and commercial research) is only realized if these data and collections are preserved. Additionally, existing collections have been assembled over many years using samples from the same sites or regions. These collections usually are much larger and more representative than collections assembled by a single expedition. Lastly, re-collection of physical samples often requires physical disturbance, which in densely populated, reclaimed, or pristine areas, could make access and collection undesirable.

WHAT ARE GEOSCIENCE DATA AND COLLECTIONS AND WHY ARE THEY IMPORTANT?

Geoscience data and collections come in many shapes and forms (Figures 1-1 and 1-2; Appendix D). Whether they are fossils, rocks, or cylindrical cores of rock, sediment, or ice, these geological materials record chapters of Earth's history.[4] Taken together, these chapters constitute a library that federal and state agencies, university researchers, and private companies use daily to understand the physical world—past, present, and future. This library provides invaluable and, in many cases, unique information with scientific, health, safety, commercial, and educational benefits, many of which are explored in this section. Each time a geological sample or piece of data is allowed to deteriorate, or is damaged, misplaced, or thrown away without assessing its merits, the information it contains and the knowledge it represents are lost. Multiplying this loss over and over again is

---

[2]Good accompanying documentation means adequate supporting information about the geoscience data and collections. What is adequate for one purpose may be inadequate for another. However, in general, documentation has to be more complete for legal or research purposes than for teaching and display purposes. Documentation includes information about age, location, depth, originator, and date acquired.

[3]In one example, the replacement cost per foot of oil industry core currently is between $550 and $1,200, compared to an average annual storage cost of $0.33 to $0.66 per foot (Emily Stoudt, ChevronTexaco, personal communication, 2001). For further discussion of this point see Montgomery (1999).

[4]Chapter 3 describes in detail the types of geologic data and collections, who collects them, and why.

FIGURE 1-1 Examples of geoscience samples and collections that provide the underpinnings for geoscience data. A) Geological outcrop. Such exposures provide a wealth of geoscience data and collections to the geoscience community. B) Rock and mineral specimens. C) Fossil fish and trilobite. D) Slabbed piece of *core* (about 4 inches across) from an industry *drill well*. E) Foraminifera microfossils. F) Tray of samples reposited at Alberta Energy and Utilities Board–Core Research Centre. The samples are drill cuttings, required to be submitted under Oil and Gas Conservation Regulation (Oil and Gas Conservation Act–Alberta) for wells that fit the criteria for this requirement. G) Sections of core from an industry drill well being unloaded for processing. H) Lower halves of typical *box* (approximately 3 feet long) of slabbed industry core. I) Microscopic fluid inclusions in potential reservoir rock for oil. J) Pictures of tapes containing data from boreholes—stored in a field computing center. K) Storage of digital audio tape (DAT) tapes containing borehole data. Original tapes and their backups are stored in the same drawer. SOURCE: A–I: ExxonMobil Upstream Research Company; F: Guenter Wellmann, Alberta Core Repository, Calgary; J and K: Phillipe Theys, Schlumberger, Ltd., Sugarland, Texas.

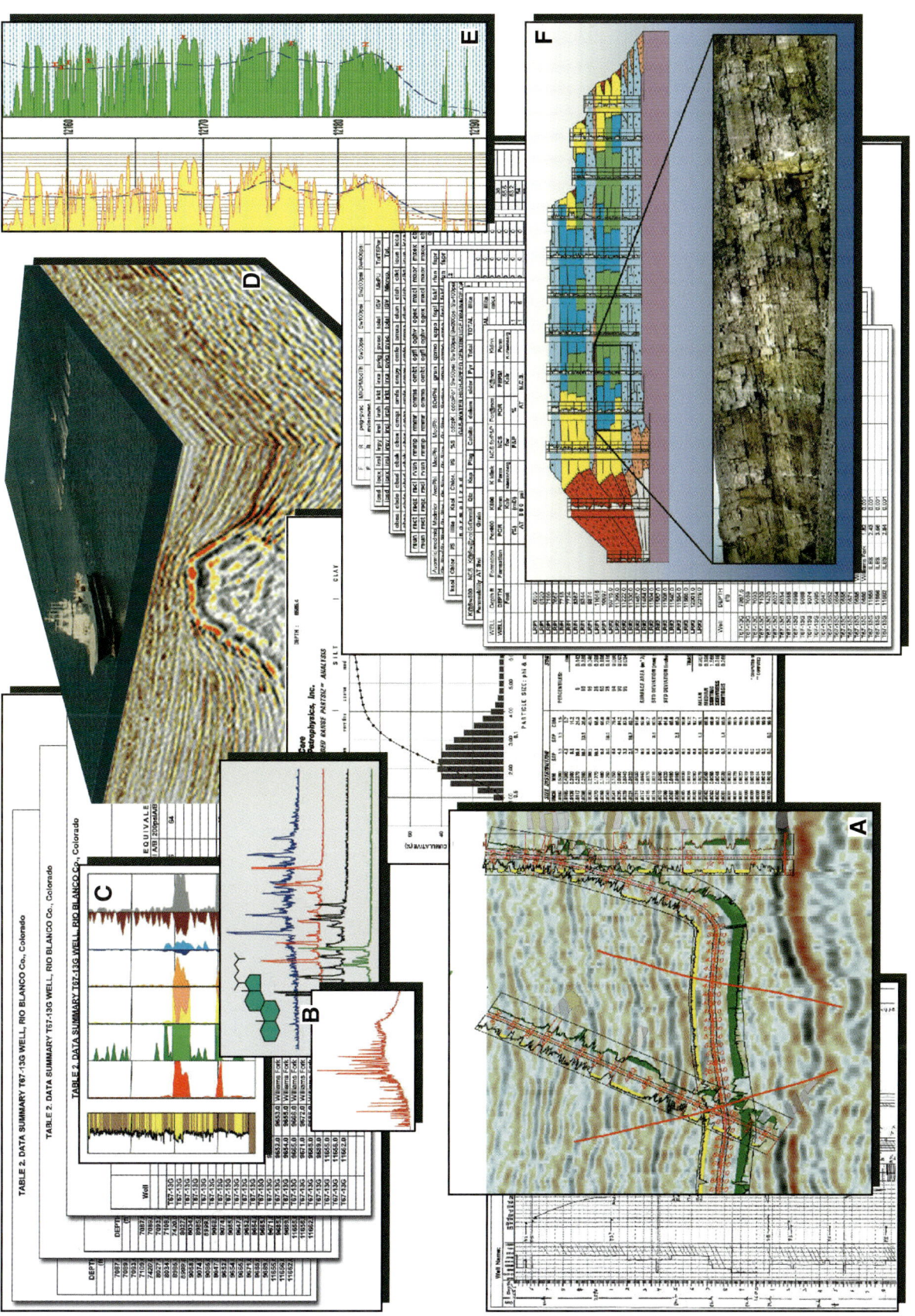

FIGURE 1-2 Examples of geoscience data. Tables and charts that underlie these images are representative of tabular geoscience data also at risk. Many of these data still reside in paper form or on old tapes. A) Two-dimensional seismic section with well *logs* plotted on the seismic data. B) Gas chromatograph plots derived from analyses of oil samples. C) Down-hole *profile* of **hydrocarbon** content of fluid inclusions. D) Three-dimensional seismic data cube. E) Well-log plot from an industry drill well showing rock type and possible hydrocarbon-rich layers. F) Outcrop of **sedimentary rocks** and a geological cross section (upper image) derived from data collected at that outcrop. SOURCE: ExxonMobil Upstream Research Company.

analogous to the destruction or loss of irreplaceable books in a library. Such losses have the potential to result in analyses, interpretations, and policies that reach incomplete, poorly supported, or even erroneous conclusions. Just as all books in any given library are not used all the time by all people, all samples and repositories of geoscience data and collections may not experience uniformly high usage. "It is a characteristic of this stuff that there are long periods of low interest and short bursts of high interest." (William Fisher, Bureau of Economic Geology, University of Texas at Austin, personal communication, 2001).

"Building large collections of crucial source material has remained a way to ensure the vitality of knowledge" (Montgomery, 1999, p. 84). New ideas almost always are built from previously collected information. "We see in the rocks what we know," says Robert Weimer, professor emeritus of the Colorado School of Mines (personal communication, 2001), and as time goes by we know more. In this way, old core can produce new knowledge. Existing geoscience data and collections may be viewed both with new eyes and with new technologies. New analytical devices or techniques may be applied to previously collected cores and other samples. New computer techniques may be applied to existing data.

Hazard Assessment

Geoscience data and collections are important resources for assessment, monitoring, and design of response strategies for many natural hazards, including volcanoes, earthquakes, landslides, and coastal erosion. Assessing the relative hazard potential of these natural phenomena relies upon one of the most fundamental axioms of geology: the present is the key to the past and the past is the key to the future. Whether determining the frequency of volcanic ash falls or landslides, the extent and rate of shoreline erosion, or the history of earthquakes in a region, it is critical to have access to geoscience data and collections that record the history of these events (Figure 1-3).

Time is critical in hazards response. Time wasted in recollecting data, tracking down lost data, or trying to upgrade existing data results in delayed response after hazardous events. Response adequacy and timeliness are directly related to data quality, quantity, and accessibility. For instance, inaccurate mine maps and inadequate site characterization have contributed to a number of coal waste impoundment failures over the past two decades (NRC, 2002, p. 29). Complete, accessible, and timely data would have saved lives and enormous sums of money. The tragedy is that many of these data undoubtedly existed but were lost, discarded, or forgotten.

Human activity requires massive amounts of energy (coal, gas, coal, nuclear, oil), minerals (aluminum, copper, iron, magnesium, iron, zinc), and water, as well as moving massive amounts of earth materials and disposing of massive amounts of waste. Screening sites for municipal, toxic, or nuclear waste disposal, designing and siting highways, bridges, dams, utility lines, and virtually every building is highly dependent upon the availability of, and access to, geoscience data and collections. Planners must have timely, complete, and well-documented information on geological formations, groundwater flow, seismic frequency and magnitude, and *geotechnical* properties of the material on which they build. Multiple core and data collections provide designers and engineers with material and datasets that are generally larger, better, and more cost-effective than can be constructed from a single new collection (assuming that building a new collection is even possible).

Basic and Applied Scientific Research

Geoscience data and collections are fundamental tools for assessing Earth's resources and for understanding Earth's geological past and the history of life. They help us address some of our most basic questions: How did Earth form? How old is Earth? How did life develop on Earth? Where did energy and mineral resources form? Why and how often do big earthquakes occur? How can we safely dispose of waste? Such investments in applied and basic research contribute to our immediate economic well-being (see Jones and Williams, 1998), as well as to the quality of our lives.

Geoscience data and collections also provide essential documentation required to address important questions of both immediate and long-term societal relevance (Allmon, 1994) (e.g., global climate change and ground water quality and availability). Geoscience data and collections provide a baseline for determining natural rates of change and estimating the frequency of natural events. Longer and more extensive records (i.e., better data and collections) result in better analyses, more accurate assessments, more definitive conclusions, and more timely responses. For example, analyses of gas trapped in bubbles in *ice cores* collected over the past decade have revolutionized scientists' thinking about possible mechanisms of climate change and the rapidity of that change. These ice core data—along with information from fossil collections, lake cores, deep-sea cores, and other sources—strengthen our predictions on the path and consequences of future changes in global *climate*. Other applications of benefit to society range from the best practices for constructing coal-waste impoundments (NRC, 2002) to the paleo-immunological implications of blood compounds preserved in dinosaur bones (Schweitzer et al., 1997). In essence, geoscience data and collections can be sources of genuine scientific discovery and real application of science for society.

Discovery, Assessment, and Enhanced Utilization of Natural Resources

Discovery, assessment, and utilization of minerals and energy resources on federal, state, and private lands are of great importance to the federal government's strategic

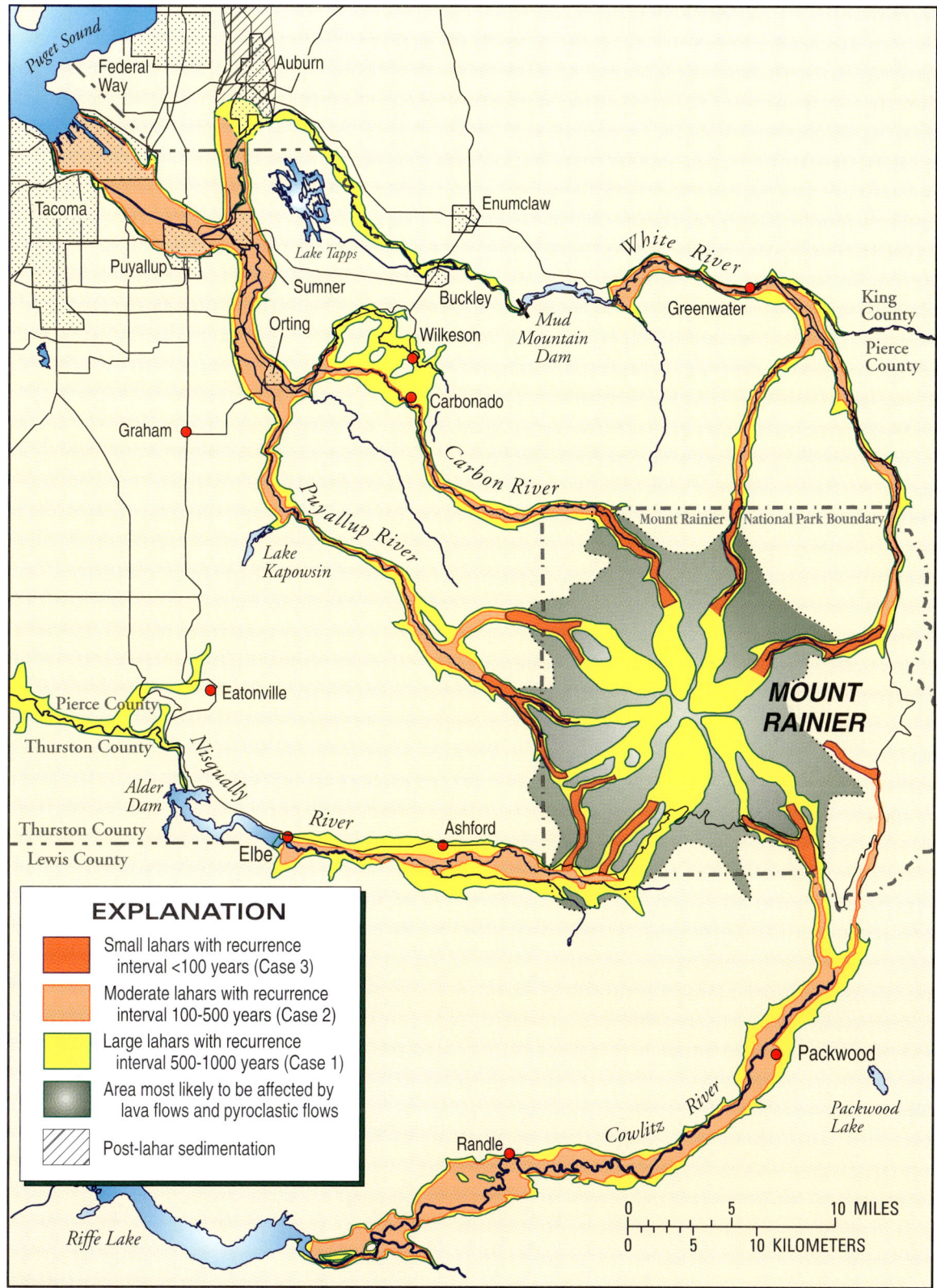

FIGURE 1-3 Hazard zones for lava flows, lahars (muddy mixtures of volcanic ash and water), and *pyroclastic* flows from Mt. Rainier, Washington. Maps such as this, which are critical for area planners and policy makers, typically are constructed from multiple types of archived geoscience data and collections. In this case, hazards assessment and predictability depend directly on completeness, accuracy, and understanding of the history of such events in the area—a history recorded in geoscience data and collections. SOURCE: Hoblitt et al., 1998.

planning. The quality, quantity, and continuity of oil and gas *reservoirs,* coal seams, ore bodies, and water *aquifers* can be verified only through a systematic examination of cores, cuttings, well logs, and seismic data (NRC, 1999a). These types of information are fundamental to the resource-assessment process, which is essential for the development of a rational energy and natural resource policy that effectively balances use, conservation, and needs.

Major advances in technology within the petroleum industry in recent years underline the importance of maintaining collections of data and samples. There now exists great interest in *field* redevelopment and enhanced characterization of known reservoirs (see NRC, 1996a, 1999a; DOE, 2002) and expansion of effort into previously inaccessible or abandoned areas. All of these new development or redevelopment efforts could bring increased domestic production and tax revenues.[5] No redevelopment program can be rationally planned without a basic understanding of the rocks themselves (e.g., through geochemical testing or hardness determination, which cannot be done without a physical sample). No minerals assessment can begin without knowing which parts have been mined previously, the remaining thickness and depth variability, the quality and value of the resource, and the geographic extent of the resource. Existing cores, cuttings, well logs, maps, and seismic data are vital to those efforts. In addition, rapid advances in geologic, geophysical, and engineering science have occurred in the computer age. With these advances it is imperative to be able to examine and analyze anew previously acquired geoscience data and collections.

Education and Public Awareness

Geoscience data and collections are important resources in training and educating the next generation of scientists, engineers, and the general public (e.g., Sidebar 1-2). The use of collections for teaching purposes is of great importance in this virtual-image, computer age. Seeing and touching the real object, whether a natural history specimen or an ash deposit in a core, is a more informative learning experience than looking at a picture or reading a description. Some features can be understood only by directly examining the rocks

---

[5]For example, DOE (2002) states that, through a DOE–University of Utah–industry partnership with a total investment of $5.8 million (of which only 38 percent came from DOE), new extraction methods were developed to bring back into production part of the previously abandoned Midway–Sunset field in Bakersfield, California. Geologists used existing core and other geoscience information to create a geologic profile of the reservoir with unprecedented detail, then used this profile to pinpoint optimal areas for new wells. Since reactivation, the field has produced more than 1 million barrels of oil (which is more than half as much oil as the lease produced between 1906 and 1985). DOE projects that application of this technology to only half of the other oil wells in the field could produce another 80 million barrels of oil, with federal tax income of more than $10 million from the oil alone.

SIDEBAR 1-2
Increased Use of Core for Research Articles in the *Bulletin of the American Association of Petroleum Geology*

It is sometimes assumed that, in the age of digital subsurface **seismic profile** data, physical cores are less important in petroleum exploration. This apparently is not the case. Articles published in the *AAPG Bulletin* that depended on core data as a basis for research and exploration were compared for two periods: those published in 1979–1981 and those published in 1996–1998 (Montgomery, 1999). Between 1979 and 1981, 38 percent of *AAPG Bulletin* articles depended on information obtained directly from core samples; between 1996 and 1998 this percentage had risen to 43 percent. In other words, usage of physical samples increased, even as digital subsurface seismic profile data became more prevalent.

---

or fossils themselves. Preparing the next generation of scientists for basic Earth science research, hazard and resource assessment, and petroleum exploration requires systematic study of the geoscience data and collections if these scientists are to fully understand the complexity of the geologic environment. And of equal importance, conveying that required knowledge to the general public demands tangible and timely information.

INVESTMENT IN AND COMMERCIAL VALUE OF GEOSCIENCE DATA AND COLLECTIONS

Billions of dollars have been invested in acquiring and building geoscience data and collections over the past 100 years. If maintained properly, those worth keeping increase in value over time because of their uniqueness and because of the information they add to more recently acquired data and collections. If not maintained properly, geoscience data and collections actually decrease in value over time and can even accrue a negative value because of the cost associated with their disposal.[6] The value of geoscience data and collections is related to their importance, the new analyses and interpretations that can be drawn from them, and the high

---

[6]For example, disposal does not require cataloging and usually does not involve special handling, but disposal does require inventorying, testing for hazardous content, packing, loading, shipping, and unloading, in addition to disposal fees. BP-Amoco and ChevronTexaco estimated disposal costs to be 35 to 50 percent of the cost of retaining and moving its materials (Jimmy Denton, BP-Amoco, personal communication, 2001; Emily Stoudt, ChevronTexaco, personal communication, 2001, respectively).

cost, in many cases, of their acquisition. Geoscience data and collections have the potential to provide solutions to future scientific, economic, and environmental issues. In effect, geoscience data and collections form the equivalent of a biography of Earth and should be thought of in much the same way. A biography with muddled dates, confused sequences of events, and missing spans of time is incomplete and inadequate at best. In the absence of high-quality, complete data, such a biography may even reach the wrong conclusions. So, too, geoscience data and collections must be of high quality and as complete as possible if scientists are to reach valid conclusions about the biography of our planet.

The U.S. Investment in Geoscience Data and Collections

Substantial investments are made to acquire data and collections. The total costs cannot be tabulated accurately, but a few examples serve to illustrate this point. The U.S. Geological Survey (USGS) Core Research Center in Lakewood, Colorado, holds cores representing nearly 235 million feet (44,508 miles) of drilling with an estimated replacement cost on the order of $10 billion (NRC, 1998).[7] The Ocean Drilling Program (ODP) has collected and archived nearly 1,400,000 feet (263 miles) of extremely costly (at least $360 per foot), difficult-to-obtain deep-sea core. During the period from 1983 to 1998, ODP activities cost the United States more than $500 million (NRC, 2000), and support continues today at approximately $46 million annually from U.S. and international sources (Frank Rack, Joint Oceanographic Institutions, personal communication, 2001). State geological surveys also hold collections that were acquired at great cost: for example, the Kentucky Geological Survey (2001) estimates the value of its holdings at $535 million. Lastly, the phase I study of the American Geological Institute's (AGI) National Geoscience Data Repository Study identified more than 100 million line-miles of seismic data that oil, gas, and mineral companies were willing to contribute to a national repository. These seismic datasets, which are no more than 25 percent of all seismic data collected in the United States since 1950, represent tens of billions of dollars of geophysical data (AGI, 1994).

Many data and collections are difficult to evaluate simply because they are unique or cannot be sampled again. For example, urban development, environmental restrictions, and other land access issues have resulted in numerous areas being closed to new drilling, sample and fossil collection, or data acquisition. Geoscience data and collections from such areas should be among those with the highest priority for preservation. For example, during the first half of the 20th century, thousands of wells were drilled in the Los Angeles Basin, among the country's most productive petroleum *basins* (see Sidebar 1-3). These previously assembled petroleum datasets are seeing a second, unanticipated use as vital information for studies linking faults with earthquakes in the basin. When preserved properly, such information is critical for thoughtfully designed urban planning studies in a variety of geologic settings.

Commercial Value of Geoscience Data and Collections

A primary goal of the President's energy policy in the coming years is to increase domestic oil, gas, and coal production (National Energy Policy Development Group, 2001). It is important, however, to view this goal through the shifting realities of the domestic oil and gas industry in particular. As large oil and gas companies commit more and more of their operating budget to deep-water offshore and non-U.S. ventures, future exploration and development of gas and oil resources on this continent increasingly will be conducted by smaller (fewer than 50 employees) independent companies (Jordan, 2000). Small and medium-sized (50 to 200 employees) independents now drill 85 percent of all new wells and account for 40 percent of oil and 65 percent of natural gas production in the lower 48 states, mostly from *marginal wells* (Jordan, 2000).

The onshore and shallow offshore areas of the United States are mature areas for petroleum production, which means that most of the major oil fields have been located, and within these fields the most easily accessible oil has been extracted. On average, only 30 to 40 percent of original oil in place (OOIP) has been recovered, while an additional 21 percent of OOIP is mobile and recoverable with known techniques, but by-passed during primary recovery and *waterflooding* (NRC, 1996a). Consequently, under the right economic conditions, future improvements in geological understanding or engineering expertise could lead to further oil production from existing fields. The original cores and cuttings, preserved in repositories, are needed to fully exploit these potentially recoverable resources. Because of trends within the petroleum industry, core repositories increasingly serve smaller companies operating in domestic fields, which leads to more efficient use of domestic resources. Increased demand for domestic oil and gas production will in turn increase demand for information about older fields, and much of this information lies in cores and cuttings—if they have been preserved.

Access to geoscience data and collections increases efficacy of companies in the exploration and development of a new field and in the redevelopment of old fields (see Sidebar 1-4, for example). These data and collections are particularly important to the independent oil and gas companies for several reasons: the cost of acquiring new data can be prohibitive for many independents; previously collected data

---

[7]Even if $10 billion were available, not all of the core holdings could be duplicated, either because the rocks or sediment no longer exist or are no longer available or because acquisition is no longer allowed. Therefore, taken as a whole, the collection represents a very expensive and in places irreplaceable investment in knowledge.

SIDEBAR 1-3
Los Angeles Basin

On January 17, 1994, the greater Los Angeles area was struck by a devastating earthquake. The magnitude 6.7 Northridge earthquake caused heavy damage in the area and more than 20,000 people were displaced from their homes, more than 9,000 were injured, and 57 died (USGS, 1996, p. 2). The Northridge earthquake has been called the most costly earthquake in the history of the United States, with damage estimates ranging from $20 billion to more than $40 billion (Eguchi, 1998; USGS, 1996).

The Northridge earthquake occurred at depth on a concealed or **blind thrust fault** that had not been recognized as a seismic hazard (Davis and Namson, 1994; Yeats and Huftile, 1995). Identification of blind thrusts, many of which have the potential to generate earthquakes, has become increasingly important in the Los Angeles area (Tsutsumi et al., 2001). For example, older data, originally collected by the petroleum industry during exploration and production activities, now are being used more effectively in the Los Angeles Basin to address these topics. Seismic reflection data acquired by the petroleum industry and made available to the public are used to locate and provide direct images of such faults (Davis and Namson, 1994; Rivero et al., 2000; Shaw and Shearer, 1999; Tsutsumi et al., 2001).

Supplementary geophysical log and sample data from oil and gas wells continue to be used to improve three-dimensional models of the sedimentary fill beneath Los Angeles (Magistrale et al., 2000). These models are used to predict the location of strong ground motions during earthquakes (Graves and Somerville, 1995), which in turn are used to improve zoning patterns and construction requirements. The damage caused by the Northridge earthquake highlights the risks faced by urban centers located in earthquake-prone areas. Better definition of seismic hazards and risks in such areas is of critical importance. The industry data used in these models could not be readily reacquired today because of restrictions associated with access, environmental concerns, and cost. This is an outstanding example of unanticipated reuse of older data to address issues that are important to all, and highlights the need to preserve geoscience data for future use.

Collapsed bridge following Northridge earthquake, span of interchange linking Antelope Valley Freeway, (California State Highway 14) and Golden State Freeway (Interstate 5), between San Fernando and Newhall. SOURCE: Jim Dewey (USGS, 1996).

> **SIDEBAR 1-4**
> **Extraction of Additional (By-passed)**
> **Oil and Gas Reserves**
>
> In many producing oil fields, primary and waterflood recovery methods produce only about 17 to 33 percent of the total original reservoir discovery. The remaining oil in the ground is a large target that tempts extraction by means of enhanced oil recovery *(EOR) techniques*. The common EOR techniques are shown in Table 1-1.
>
> Chemical, gas injection, and thermal recovery are the most common EOR techniques. These methods rely on generating increased reservoir pressure from the injection wells toward producing wells, as well as changing properties of the hydrocarbons and interface between the rock grains or particles and hydrocarbons. With successful EOR projects in combination with primary and waterflood reservoir extraction techniques, as much as 50 to 70 percent of the OOIP can be recovered.
>
> Other significant sources of hydrocarbons occur in nonproductive intervals between or within productive wells and in dry holes. These by-passed intervals often were interpreted originally as water-bearing, having inherent low *permeability*, or very damaged by drilling fluids, and therefore, not of commercial quality. Poorly swept oil reservoirs, or zones with oil mobility problems also have resulted in recoverable oil being left in place. In recent years, geologists have applied new methods for reservoir characterization to old data to revitalize many older oil fields. The challenge to today's petroleum geologist is to explore within the old field by using existing data. Numerous examples illustrate successful commercial production where oil is derived from *by-passed pay zones* (Sneider and Sneider, 2001; see also Elmworth field example in Sidebar 1-5).

can be used to target likely areas of new activity; and an existing field can be redeveloped more cost-effectively and efficiently with an understanding of the properties of the rocks, data which come directly from geoscience data and collections (e.g., DOE, 2002). The benefits of well-documented data and core collections to independent oil and gas companies also include less wasted staff time in searching for data, better estimates on long-term costs associated with re-evaluation of a field and, ultimately, shorter exploration and development time, and better production.

Simply put, access to geoscience data and collections can mean the difference between by-passing an extractable resource or not, or worse yet, attempting to extract a resource that is not economical because critical data were missing, which led to an erroneous conclusion and waste of time and money. Access to geoscience data and collections also benefits states and the federal government. Secondary use of old data has added significant resources to the nation's oil and gas supply (NRC, 1996a). Furthermore, approximately 50 percent of the remaining, untapped technically and economically recoverable crude oil and gas resources are located on federal lands (DOE, 1999), and the most recent compilations of annual oil and gas royalties and tax revenue from state lands in Alaska, California, and Wyoming alone exceed $2 billion.[8] Sidebar 1-5 illustrates the benefits of reuse of old data for industry and residents of the northeastern United States and Canada.

Given all these benefits, why don't independent oil and gas companies maintain their own geoscience data and collections repositories? Many of the smaller companies do, but this is because their total holdings are small and occupy minimal space. As a company grows, the reference library of geoscience data and collections also grows. Space becomes less available until something has to be done to make room for new collections. Unfortunately, the one-time cost of handling this problem can be prohibitive to an independent operator when a company goes out of business or is bought by or merged with another company. This is where access to a repository of accumulated geoscience data and collections is so critical.

An appropriate analogy can be found within the legal profession. A small firm with a limited regional practice typically can maintain most of its legal reference collections (i.e., in its law library) onsite. But as the firm grows and the regions and case types expand, the need for additional references also grows to a point beyond that which would fit into the firm's available space. No one questions the need for access to pertinent legal opinions and precedents. Fortunately, larger libraries (mostly public) are available, and information can be copied, shipped, or acquired directly onsite. The analogy holds for geoscience data and collections, some of which record the history of an oil field—a small volume in the library of Earth's history. With growth of a petroleum company, the number of fields and the variety of challenges increase, necessitating a larger library of cores, cuttings, etc. to properly evaluate the field. Unfortunately, regional geoscience data and collections libraries are comparatively few and many either are filled to capacity (or beyond) (see, for example, Tables 2-3 a and b) or limit the types and amounts of geoscience data and collections they are willing to take (for example, the USGS Core Research Center, Sidebar 3-2). This necessitates a burdensome amount of time spent

---

[8]SOURCE: Jim Stouffer, Alaska Department of Natural Resources, personal communication, 2001; David W. Brown, State Lands Commission, California, personal communication, 2001; Harold Kemp, Wyoming Office of State Lands and Investments, personal communication, 2001; Randy Bolles, Wyoming Department of Revenue, personal communication, 2001.

TABLE 1-1 Techniques for Recovering Oil Remaining after Primary and Waterflood Recovery

EOR Technique	Description
Chemical flooding:	
Alkaline	Inorganic alkaline chemicals (e.g., sodium carbonate) are added to injected brine to raise the pH, which in turn reacts with the acidic portion of the crude oil to produce in-situ surfactants. The in-situ surfactants recover additional oil by reduction of the interfacial tension between formation water and oil changing the wetability thereby releasing more oil.
Polymer	A small amount of polymer (e.g., polysaccharide or polyacrylamide) is added to injected brine to increase its viscosity and reduce its mobility thereby increasing the sweep of the reservoir rock and hydrocarbons. Polymer flooding is the most widely used of the chemical recovery process because of its low cost.
Surfactant/polymer	Sequential injection of several small volumes (or slugs) of chemicals into the reservoir to attain increase oil recovery. In a typical flood, surfactants are injected followed by a polymer slug and thin brine. This method is expensive but has a high hydrocarbon ultimate recovery.
Gas injection:	
Carbon dioxide	CO_2 reduces crude oil viscosity, mixes with and swells the crude oil plus provides a gas drive, all of which increases oil recovery to recover incremental oil left after waterflooding.
Hydrocarbon miscible	Injected fluid that is miscible with crude oil (e.g., methane) drives hydrocarbons to producing wells. This type of flood may use lean hydrocarbon gas or liquefied petroleum gas. The injected hydrocarbon fluid promotes the increased recovery of crude oil by eliminating of interfacial forces.
Nitrogen and/or flue gas	Nitrogen injected under pressure into a reservoir mixes with the crude oil and drives the hydrocarbon mixture to producing wells.
Thermal:	
Steam	Steam generated at the surface is injected into a reservoir containing viscous crude oil or tar. Heat is transferred to the crude or tar, which lowers its viscosity, improves its mobility, reduces capillary forces, and expands the hydrocarbons and may also distill the hydrocarbons, producing light hydrocarbons components.
Hot water	A thermal EOR technique in which hot water is generated at the surface and injected to heat a viscous crude oil or tar. See the steam flooding for the overall process.
In-situ combustion or fireflooding	Heating hydrocarbons within the reservoir recovers viscous crude oil or tar. Heat is generated within the reservoir by injecting oxygen or air and setting part of the hydrocarbons on fire. Some of the in-situ hydrocarbons are burned to generate heat and the heat mobilizes the remaining crude or tar, which is recovered in production wells.
Microbial drive:	Method involving the injection of microorganisms (e.g., bacteria) into a reservoir, which interact with crude producing several EOR compounds that increase recoverable reserves.

SOURCE: Adapted from Stocur, 1986.

seeking samples and information from other, decentralized sources, with no guarantee that they exist.

USERS AND BENEFICIARIES OF GEOSCIENCE DATA AND COLLECTIONS

Virtually everyone in the United States benefits in one way or another from geoscience data and collections (Table 1-2). Benefits vary—from the satisfaction of a backyard paleontologist who takes a latest find to an expert for identification or collaboration, to the entertainment and education of the millions of people who visit the Smithsonian Institution's rock, mineral, and dinosaur exhibits every year, to the obvious benefits to society of a diverse, effectively utilized, national energy supply, to the comforts and conveniences of modern society afforded by Earth's resources used in every home, office, and transportation vehicle.

The impact of work performed by users given in Table 1-2 is felt in the availability and price-competitiveness of domestic fuels, in the safety of our water supply, in the stability

> **SIDEBAR 1-5**
> **Elmworth Gas Field**
>
> Residences and businesses in northeastern United States and eastern Canada enjoy power generated from natural gas that comes in large part from the Elmworth field in western Canada. This ***supergiant*** field was discovered in 1976, and gas delivery to the east began in the early 1980s. The field now has more than 3,000 producing wells and is still under development. Elmworth field covers about 5,000 square miles and contains more than 20 trillion cubic feet of recoverable gas and about 2 billion barrels of oil and condensate in 15 sandstone and conglomerate reservoirs. Discovery or, more appropriately, re-discovery of the Elmworth supergiant field provides an excellent case study for the use of existing data.
>
> A small team of U.S. and Canadian geoscientists and engineers discovered significant by-passed gas pays during a 1975–1976 regional study of several hundred well logs, cuttings, and production tests from dry holes drilled in the late 1940s to the early 1960s. The old wells originally were drilled for hydrocarbons in carbonate reefs, which were below the newly identified by-passed pay zones. The geoscience and engineering data were made available through the Alberta Energy Board's Core Research Centre in Calgary (see Sidebar 4-1). In excess of 425,000 meters (1,394,354 feet) of cuttings were studied from almost 1,000 wells in the area. Of the wells analyzed, 61 were found to be hydrocarbon-bearing and within the field itself.
>
> The team used new ***petrophysical*** analysis techniques to find the pay zones and utilized new completion methods for extraction. Without the Alberta provincial facilities' systematic storage of the geoscience-engineering data, it is doubtful that the originally overlooked reservoirs would have been found in such a short time, if at all.
>
> SOURCE: Masters, 1984.

of our dams, roads, tunnels, and buildings, and in debates over environmental change and the effect of human actions on such change.

Geoscience data and collections are available to the public in every state in the United States. Visits to various facilities that hold geoscience data and collections range from a few per month to many hundreds per year (e.g., the Kansas Geological Survey [KGS] received 606 visitors to its cuttings collection alone in the first seven months of 2001 [M. Lee Allison, KGS, personal communication, 2001]). Most of the organizations the committee surveyed (Appendix B) predict that their usage patterns will remain stable or will grow, especially with enhancement of Internet technologies that offer increased access to knowledge (i.e., better metadata) about the geoscience data and collections.

Unanticipated Benefits

Perhaps one of the greatest values of geoscience data and collections is that most have uses unimagined by the original collectors. Changes in significance of a collection may occur over time as a result of human activity and natural geological and biological processes. As scientific ideas advance, new concepts within a discipline—or in entirely new disciplines—often emerge (see Sidebar 1-6). New analytical technologies elicit different and often better information from previously analyzed collections. These unanticipated uses have been extremely important, particularly when re-collection is impossible or not feasible with the time and resources available.

Data and samples the petroleum industry collected have been used later in unanticipated ways to address hazards in a number of instances. For example, core and seismic data have been used to improve evaluation of earthquake hazards in urban areas (for the Los Angeles Basin see Sidebar 1-3).

TABLE 1-2 Users and Beneficiaries of Geoscience Data

Users and Beneficiaries of Geoscience Data and Collections
Civil engineers
Climate researchers
Construction industry personnel
Defense industry personnel
Educators and students
Emergency preparedness personnel
Environmental engineers and scientists
Farmers and ranchers
Foresters
Hydrologists
Insurance industry
International commodity traders
Landowners and home-use owners
Lawyers
Oceanographers
State and federal policy makers, regulators, and agencies
The energy industry
The minerals industry
Urban planners

SOURCE: Responses to committee questionnaire (Appendix B).

SIDEBAR 1-6
A New Use of Cuttings:
Analysis of Fluid Inclusions

Fluid inclusions are micron-sized liquid- or gas-filled cavities that occur in many rocks. They are formed when minerals and **cements** crystallize, trapping samples of **interstitial** fluids present at the time of their formation. The composition of fluid inclusions is not altered by removal from the subsurface, nor are their contents modified by storage over time.

Some fluid inclusions are large enough to be seen with a microscope. Commonly, however, inclusions are very small and cannot be resolved by optical methods. Energy-industry scientists have developed techniques to analyze these small amounts of trapped fluids. Down-hole profiles derived from fluid-inclusion analyses can be used to determine zones of hydrocarbon migration, proximity to potential reservoirs and, in some cases **hydrocarbon contacts.** In a representative example, cuttings from a well drilled in 1983, long before these techniques were developed, were analyzed in 1999. The hydrocarbon content of the fluid inclusions led to a completely different reappraisal of the source area. Drill cuttings, many of which contain fluid inclusions, are proving to hold a wealth of under-utilized information, underscoring the potential benefit of long-term preservation of such material.

SOURCE: ExxonMobil Upstream Research Company, personal communication, 2001.

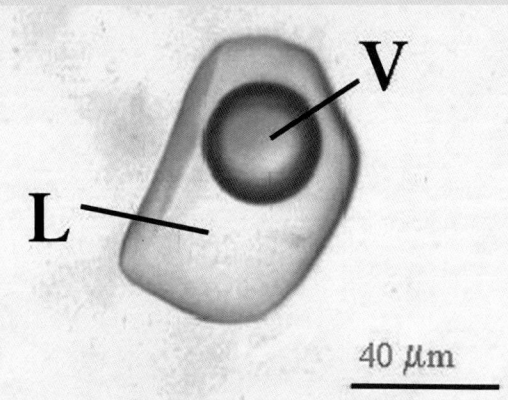

Fluid inclusion (L) containing a bubble of gas (or vacuole, V), all within a single crystal from a cutting. New technologies now allow extraction of previously unimagined information from Earth's library of rock. SOURCE: ExxonMobil Upstream Research Company.

Such evaluations are critical to urban planners and designers of construction projects in earthquake-prone regions. In 2001, core data were critical in averting a further disaster when natural gas and brine escaped from an underground storage facility in Kansas (see start of Executive Summary, and Figure 1-4).

Another example of unanticipated use comes from ice cores. Sampling techniques developed years after some cores were taken have allowed climate researchers to examine changes in the carbon dioxide (CO_2) content of the atmosphere (see Sidebar 1-7) from as long as 420,000 years ago. In Virginia, maps of mine locations have been used to assist in emergency response planning should accidents happen in old mines. Law enforcement officials in North Carolina used similar maps to search for fugitives from justice. Geoscience data even have been used for genealogical research through old lease and mining data.

FIGURE 1-4 Flames from Hutchinson, Kansas, January 17, 2001. This fire and a fatal one in Hutchinson the next day originated when natural gas migrated through rocks from a storage facility 8 miles away (see the first page of the Executive Summary for a fuller account). SOURCE: Hutchinson, Kansas Fire Department.

SIDEBAR 1-7
Ice Core Reuse

Ice cores are cylindrical sub-surface samples of glacier ice. These samples have been collected from the Antarctic and Greenland ice sheets since the early 1960s. Most U.S. cores are housed at the National Ice Core Laboratory at the Denver Federal Center in Lakewood, Colorado (see Sidebar 2-11). An important characteristic of ice cores is that they contain old air (Alley, 2000)—air trapped when the ice formed many years earlier. The deeper the origin of the core, the older the air. Near the base of the Antarctic Ice Sheet, at depths of more than 4 kilometers (2.49 miles), the trapped air bubbles are older than 400,000 years.

This old air is currently of great societal and scientific interest because it carries a record of past levels of atmospheric CO_2. For example, a central piece of information in the global-warming debate is the comparative magnitude of pre-industrial atmospheric CO_2 levels and modern values. Looking further back in time, the variation of CO_2 through glacial cycles (each cycle lasting about 100,000 years) gives clues to driving forces behind global climate change, and whether or not industrialization has affected any of these driving forces.

Techniques to measure CO_2 from bubbles within ice cores were developed in the early 1980s—a decade after the original long cores were collected at Camp Century, Greenland (1963–1966), and Byrd Station, Antarctica (1968). Fortunately, the cores were preserved in anticipation of improved analytical techniques. The results revealed atmospheric CO_2 levels for the last 30,000 to 50,000 years for the first time. Levels of CO_2 in northern and southern hemispheres during the last glaciation were shown to be roughly half the modern values. Large changes in the ***biosphere*** were likely responsible for the substantially reduced CO_2 levels during the last glaciation (Bradley, 1985). Such large changes in a gas so important in the global energy balance have profound implications for hypotheses of climate change.

Since the initial discovery of atmospheric CO_2 levels during the last glacial period, the measurement of CO_2 concentration has become routine on new ice cores. In the case of the Vostok Core from Antarctica, the record of CO_2 has now been extracted back to 420,000 years before present. In general, modern ice cores provide a substantial amount of paleoclimatic information. The existence of the National Ice Core Laboratory (NICL) is a reflection of the value now placed on ice-core data and the understanding that potential discoveries await in existing cores. This facility is jointly operated by the USGS and NSF's Office of Polar Programs. Occasionally, NICL offers old cores for destructive analysis if, for example, duplicate cores exist. This practice allows scientists to develop new techniques, such as those to analyze CO_2 levels, without fear of wasting unique or expensive samples. Ice-core science thus progresses even from cores that have no further life at the repository.

Close-up of an ice core from the Greenland Ice Sheet Project. After an ice core is cleaned, it is sawed in half lengthwise to reveal features like these seen here. The bands are formed by individual years of snow accumulation; the core in this photograph contains a sampling of 4 or 5 years of ancient atmospheric conditions. The core was collected from a depth of 1,850 meters (6,070 feet). The age of ice at that depth is approximately 16,750 years. The width of the ice in the picture is 5.2 inches. SOURCE: Geoffrey Hargreaves, NICL.

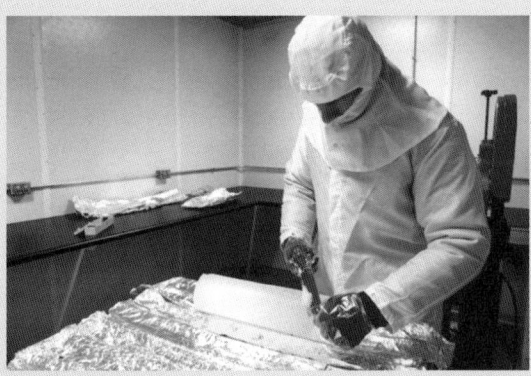

Scientist cleaning a piece of ice core in the cold clean room.
SOURCE: Geoffrey Hargreaves, NICL.

2

Nature of the Challenge

The volume and diversity of geoscience data and collections are enormous. Although the volume of digital data also has grown tremendously in past decades, technological advances have reduced the actual physical space required to store and handle this increased volume (NSF/ONR, 2001). Such is not the case with the physical geoscience collections and media containing data that are the primary focus of this committee.[1] Geoscience data and collections are both physical and digital, consequently they occupy space. This chapter contains assessments of the current volume of geoscience data and collections in the United States, the available space, the factors placing them at risk of loss, and priorities for what to retain.

VOLUME OF GEOSCIENCE DATA AND COLLECTIONS

Accurate assessment of the total volume of geoscience data and collections in the United States has been a challenge primarily because of inaccurate or insufficient metadata about geoscience materials. A second challenge is estimating how much material one entity will donate to another. Unanticipated donations have filled repositories much faster than expected, which has seriously challenged their management and their ability to accept further donations. The numbers in Table 2-1 represent estimates of total volume of geoscience data and collections in the United States. They are taken from recently published reports and from testimony and surveys the committee received. These estimates are not based on a complete inventory of information, because none yet exists. Therefore, the estimates in Table 2-1 should be viewed as minimum figures,[2] in some cases perhaps too low by as much as an order of magnitude. In addition, similar surveys (by this committee and other entities) are voluntary, and responses never total 100 percent. For example, state geological survey data in Table 2-1 represent data for 39 of the 50 states; responses to the AGI (1997) survey did not include all petroleum companies; and a Canadian Society of Petroleum Geologists (CSPG) survey drew responses from 62 of 360 inquiries (CSPG, 2001). Furthermore, there are no authoritative estimates at all for certain types of data and collections, such as the number of fossils or mineral specimens held by industry or individuals. Indeed, many repositories and companies have imprecise estimates of their own holdings. For example, the CSPG survey indicated that 52 percent of respondents did not maintain an organized database of their holdings (CSPG, 2001).

These limitations notwithstanding, it is clear that the nation's geoscience data and collections comprise large volumes. There are more than 100 million fossil specimens; more than 8 million boxes of core—containing more than 80 million feet of rock and sediment (or more than 15,000 miles, the equivalent of drilling more than twice the way though the Earth) (see Figure 2-1); more than 10 million boxes of cuttings; more than 40 million well logs; and more than 350 million line-miles of seismic data (or the equivalent of 140,000 times around the Earth) (see footnotes in Table 2-1). The size and scope of these numbers may be grasped by comparing them with the similar, but more familiar issue confronting libraries; the nation's research libraries, for example, collectively contain an estimated 400 million books (ARL, 2000).

Despite the large volume of geoscience data in the United States, some portion is in immediate danger of being lost

---

[1] Digital information about the physical collections (e.g., number of cores, intervals cored, locations, ages, images) are essential in the search for available data. However, these data about the data (metadata) can never take the place of the original because new or enhanced techniques typically cannot be applied to images and information (see for instance Sidebars 1-6 and 1-7).

[2] The committee decided to be conservative in its estimates of figures presented in this document. Thus, these (and other) numbers very likely represent the lowest in a range.

TABLE 2-1 Minimum Estimates of the Volume of Geoscience Data and Collections in the United States

	Units	Industry	State Surveys*	Other[†]	Academia	Museums	Individuals	Total
Collections								
Core (rock/sediment)	Boxes or equivalent[a]	3,500,000[b]	4,167,000[c]	334,500[d]	14,000[c]			8,015,715
Cuttings	Boxes	8,750,000[e]	1,600,000[f]	52,000[g]				10,402,000
Thin sections	Slides	105,000[e]	420,000[f]	122,000[h]				647,000
Washed residues				180,000[i]				180,000
Other well records			2,000,000[f]	45,000[j]				2,045,000
Fossils	Specimens	10,000,000[k]	235,000[f]	1,200,000[l]	23,000,000[m]	58,500,000[l]	30,000,000[k]	122,935,000
Minerals/rocks	Specimens		128,000[f]		100,000[f]	600,000[f]		828,000
Core (ice)	Tubes			14,500[n]				14,500
Data								
Seismic (2-d)	Line-miles	355,250,000[e]		1,732,000[c]	38,000[o]			357,020,300
Seismic (3-d)	Square miles	90,000[p]		160,000[q]				249,849
Velocity surveys	Paper and digital	87,500[e]						87,500
Well logs	Paper, films, fiche, tapes	24,850,000[e]	3,500,000[c]	17,500,000[o]	173,000[o]			46,021,700
Scout tickets	Fiche and paper	8,750,000[e]	1,781,000[c]	11,129,000[o]	300,500[o]			21,960,350
Geochemical analyses	Paper	1,750,000[e]						1,750,000

*State and oil gas commissions and state departments of conservation.
[†]Includes federal government.
[a]Box = 10 feet of core.
[b]AGI (1999). This is a minimum estimate.
[c]AGI (1997) and NRC committee survey responses, 2001.
[d]USGS (120,000), ODP/DSDP (197,905), USACE (Alaska)—NRC committee survey responses, 2001, DOE (Yucca Mtn), and AGI (1997).
[e]AGI (1999) reported 3.5 million boxes of core in all major oil companies; it then reported 1 million boxes potentially available for donation, together with other types of data. We assumed that this ratio of 3.5 total to donatable material was valid for all other types of data listed in the AGI report.
[f]NRC committee survey responses, 2001.
[g]USGS; NRC committee survey response, 2001.
[h]USGS and ODP; NRC committee survey resonses, 2001.
[i]ODP; NRC committee survey response, 2001.
[j]CWSR, LABSDC; NRC committee survey responses, 2001 (see Appendix F for acronyms).
[k]Allmon (1997, 2000).
[l]White and Allmon (2000).
[m]Includes university museums.
[n]NICL; NRC committee survey responses, 2001.
[o]DERL, HGRC, PII, IOGS, JLL, OCGSL, RELI, BELI, CGSI, EII, GP, MEL, OILF, OILW, SSPLA (AGI 1997).
[p]Elwyn Griffiths, ExxonMobil, personal communication, 2001. Estimate is for areas with known 3D coverage. An uncertainty of 20 percent is estimated due to an unknown amount of data from areas with more than one seismic survey.
[q]Includes offshore industry data copies submitted to MMS; Dellagiarino et al. (2000) (16,094); 1,667 other, AGI (1997).

FIGURE 2-1 1,000 feet (333 boxes) of rock core laid out inside the Bureau of Economic Geology Core Facility, University of Texas at Austin. These rows represent data from four wells. Given the average increase in core and cuttings holdings annually, these 333 boxes represent approximately 2 months of average growth (see discussion in Sidebar 3-4). SOURCE: David M. Stephens, Bureau of Economic Geology, The University of Texas at Austin.

because of inadequate space or incentive to retain those worth keeping. Estimating the portion at risk is even more challenging than estimating the total volume. However, the committee estimates that half of all data and collections held by industry, and at least 25 percent of collections held by individuals, academia, and government are endangered. This means that millions of items—specimens, boxes of core and cuttings, tapes, fossils, and paper documents—are in peril of being lost. A single facility capable of holding just these endangered materials (i.e., 2 million boxes of core, 4 million boxes of cuttings, 12 million well logs, 150 million line-miles of seismic data, 10 million fossils, with no room for additional samples) would have to be at least 20 times as large as the current USGS Core Research Center in Lakewood, Colorado.

The primary sources of potentially available (and there-

SIDEBAR 2-1
Findings of the American Geological Institute (AGI) in 1997

Large amounts of geoscience data and materials already have been identified and are in need of storage and curation. As part of a multi-phase study, AGI surveyed private industry participating in geologic activities, largely the major independent petroleum and mining companies. Their 1997 report illustrates that the items in the following table could be expected as a minimum initial contribution of geoscience data and collections from the natural resources industries.

Geoscience Data Available for Transfer from Natural Resources Industries to the Public Domain in 1997:

Cores	10,000,000 linear feet (about 1 million boxes)
Cuttings	2,500,000 boxes
Thin sections	30,000 slides
Seismic (hard copy)	1,500,000 line-miles
Seismic (films)	1,000,000 films
Seismic (digital)	100,000,000 line-miles
Related data	25,000 velocity surveys
Well logs (paper)	5,000,000 logs
Well logs (fiche)	1,500,000 fiche cards
Well logs (digital)	600,000 tapes
Scout tickets	2,500,000 fiche and paper
Geochemical analyses	50,000 paper

SOURCE: AGI, 1997.

TABLE 2-2 Examples of Transfer of Cores from Corporate-Owned Repositories to State Geological Surveys

Donating Company	Year Donated	Receiving Organization	Number of Boxes Donated
Shell Oil Co.[a]	1994	BEG (see [a]); BEG shipped 2,134 boxes of core to NMBGMR (see[b])	400,000[d]
Burlington Resources[b]	1997	NMBGMR	535
Altura Energy Ltd.[b]	1999	NMBGMR	5,502
Amoco Production[c]	1999	Geological surveys of Oklahoma, Texas, Utah, New Mexico (NMBGMR)	6,000
El Paso Energy/Sonat[b]	1999	NMBGMR	4,292
Altura Energy Ltd.[a]	2000	BEG[e]	85,000
BP Amoco	2000	Kansas Geological Survey	8,258

[a]SOURCE: George Bush and Scott Tinker, Bureau of Economic Geology (BEG), University of Texas at Austin.
[b]SOURCE: Ron Broadhead, New Mexico Bureau of Geology and Mineral Resources (NMBGMR).
[c]SOURCE: Jimmy Denton, BP Amoco, Tulsa.
[d]Plus 50,000 boxes of cuttings.
[e]At the time this document was going to press, a major oil company was in negotiation with the BEG and another state geological survey involving a donation of similar size to the Shell Oil Co. donation of 1994.

fore threatened if no other facility can take them) geoscience data and collections are major oil companies, independent petroleum producers, and mineral extraction companies (AGI, 1997). An American Geological Institute survey (AGI, 1997) estimated how much material these groups would consider contributing to the public domain if facilities existed to receive the information. Sidebar 2-1 summarizes these results. Table 2-2 shows examples of donations of core from industry to the public sector since 1994.

A NATIONAL SHORTAGE OF SPACE

Although it is difficult to quantify the amount of space available in the nation's repositories, many are essentially at or near capacity. Repository managers therefore are refusing to accept new data and collections because they simply do not have enough space in the repository. Figure 2-2 illustrates the amount of space available at state geological survey repositories around the United States, and Tables 2-3a and 2-3b summarize, respectively, the available space at state geological surveys and at other entities across the nation. Of the 35 responding state geological surveys, nearly two-thirds have 10 percent or less available space, and nearly one-quarter are entirely full. At least one-third of the state geological surveys listed in Table 2-3a have been forced to turn away geologic materials, and more than three-quarters of them could not add new space. The cross-section of other geoscience repositories around the country (Table 2-3b) reveals similarly low amounts of available space.

The following situation is typical: because of limited space, a repository can accept core and other physical data only if it discards a similar volume. The result is that every time something is added, something else must be removed. The repository can apply its own set of criteria, but without formal protocols to set priorities, valuable data and collections may be at risk from the limited assessment of a single individual. Repository managers may try to preserve geoscience data and collections in other ways, such as by offering discarded material to other qualified organizations or using it in other ways (e.g., student study sets).

Because of the vagaries in knowing how much new material might be offered, repository administrators often have difficulty estimating how quickly remaining space will fill up. Examples include the state geological surveys of Kansas, Kentucky, and Ohio, all of which have added new space

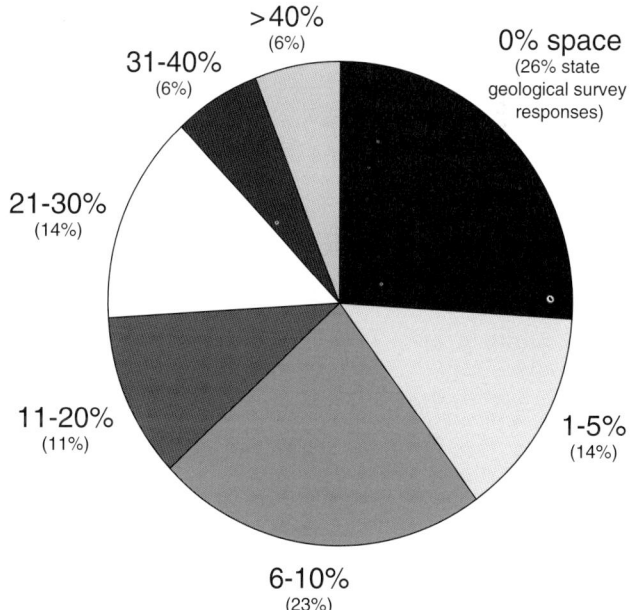

FIGURE 2-2 Percentage of available space for cores and samples at state geological surveys (based on data in Table 2-3a, which was compiled from 35 responses to the committee's questionnaire). Nearly two-thirds (63 percent) of the state geological surveys that responded to the committee's questionnaire reported that they have 10 percent or less remaining space for geoscience data and collections.

TABLE 2-3a Available Space and Refusal of Samples at 35 State Geological Surveys

State Geological Survey	% Space Available	Refused Samples?
Alabama	10	N
Alaska	30[a]	N[b]
Arizona	10	Y[c]
Connecticut	0	Y
Delaware	10	Y[c]
Florida	0	
Hawaii	50[d]	N
Illinois	2	Y[e]
Indiana	0	Y[c]
Iowa	25	
Kansas	10	N
Kentucky	<5[f]	N
Louisiana	0	Y
Maine	0	
Michigan	15	
Minnesota	25	
Missouri	20	Y[c]
Montana	0	Y
Nebraska	25	N
Nevada	50[g]	Y[h]
New Hampshire	<0[i]	Y
New Mexico	12	Y[j]
New York	<5[f]	N[k]
North Carolina	40	N
North Dakota	30	N
Ohio	16	N
Oregon	10	N
South Dakota	40	
Tennessee	0	
Texas	10	
Utah	10	
Virginia	5	Y
West Virginia	<10	
Wisconsin	0	Y
Wyoming	<5[f]	N

[a]Large cargo transport containers (**CONNEX containers**) with shelves provide additional space for an already full repository.

[b]"but does not actively try to obtain specimens."

[c]"have to be selective about what to accept."

[d]"In terms of percent, it's hard to say. We can always store cuttings and core samples."

[e]"accept some collections to save them from disposal."

[f]"almost no space" or "very little" is interpreted to mean less than 5%.

[g]Ocean-transport containers have provided additional space to an already full repository.

[h]"accept only a small representative set."

[i]"...we are losing space, not gaining."

[j]refused very large donations of core unless accompanied by money to build new core storage facilities.

[k]NYC core to Hofstra University, which is full.

SOURCE: Questionnaire by the Committee on Preservation of Geoscience Data and Collections.

TABLE 2-3b Repository Space for Long-term Archiving of Geoscience Data and Collections at a Cross-section of Non-State Geological Facilities in the United States

Facility Name	% Space Available for Long-term Storage
Smithsonian[a]	<15
U.S. Army Corps of Engineers[b]	0
U.S. Geological Survey Core Research Center	10
Ocean Drilling Program	11
C & M Storage Inc.[c]	12
LA County Museum of Natural History	0
University of Rhode Island	30
California Well Sample Repository- Bakersfield	10
Denver Earth Resources Library	0
National Ice Core Laboratory[d]	10
Los Angeles Basin Subsurface Data Center	33
National Lacustrine Core repository	75

[a]Includes space available for departments of paleobiology and mineral sciences at the National Museum of Natural History, Museum Support Center and Garber facility.

[b]Percentage based on requirements in the U.S. Army Corps of Engineers document ER 1110-1-1803 that calls for retention of core for 5 years or longer if in litigation. In practice, most cores are kept through the construction and litigation phases, which typically span about 10 years (Michael Klosterman, USACE, personal communication, 2001). Therefore, the Army Corps has no policy to retain long-term data.

[c]C&M Storage Inc. has the potential to expand a further 64% above current capacity with the addition of new buildings on land they own (Robert Shafer, C&M Storage, personal communication, 2001).

[d]Compactorized shelving is planned, which will increase in available storage space.

SOURCE: Questionnaire by the Committee on Preservation of Geoscience Data and Collections (see Appendices B and C).

since 1990, yet none have more than 16 percent remaining. The Ocean Drilling Program, which is in an enviable position of knowing how much new material will be generated, estimates that it will require new space in 2004 (Frank Rack, Joint Oceanographic Institutions, personal communication, 2001). This is sufficient lead-time to plan to accommodate the material.

In contrast, a single donation of material from a major oil company (e.g., Table 2-2) easily has the potential to push a public facility beyond its limit unless the prospective donor also is willing to donate the building containing the materials, such as the case of Shell Oil's Midland, Texas, facility, which was donated to the Bureau of Economic Geology, University of Texas (see Sidebar 2-2). Innovative public-private partnerships such as the Shell donation exemplify the importance of providing incentives for donation of geoscience data and collections to public entities.

Few managers of repositories currently view themselves as having the luxury to plan for regular additional growth; almost all major growth occurs unexpectedly (or with minimal advance notice) via donation of collections from other

> **SIDEBAR 2-2**
> **Shell Oil's Donation of Geoscience Data:**
> **A Success Story in Texas**
>
> Shell Oil's transfer of its core facility at Midland, Texas, to the Bureau of Economic Geology at the University of Texas illustrates one model for transferring geoscience data and collections from the private sector (Montgomery, 1999). From 1993 to 1997, Shell analyzed options for the geoscience data in its seven repositories. Shell determined that the $1 million annual maintenance costs were a significant financial burden to the company for areas it no longer considered central to its business, yet it also recognized the value of the cores and data. Shell chose to resolve the issue through an innovative public–private partnership.
>
> Shell deeded a collection of 2.2 million linear feet of core (450,000 boxes) to the University of Texas, together with its warehouse. Altura Energy, Ltd. donated money for the storage building. Shell also provided the university with a $1.3 million endowment (in 1995) to help cover annual operating costs. The amount of money donated was estimated by the BEG to be the amount necessary to begin an endowment to run the facility. The company retains full access to the material under the same arrangements as all others wishing to access the material. Initial operating costs were offset by a grant from the Department of Energy, which allowed the Shell endowment to increase to more than $3 million in 2002 (Douglas Ratcliff, BEG, personal communication, 2002). Although final details remain to be worked out with the federal side, these cores and data are now available in the public domain for the first time and can be used for scientific, educational, and commercial purposes.

repositories. Most repositories, therefore, appear to be constantly on the edge of moderate to severe overcrowding. There are few acceptable ways to decrease space without losing useful geoscience data and collections. As space fills over time, repository managers are forced to turn away other geoscience data and collections. Acquiring additional space for the repository can alleviate the problem, but new space typically is small, so relief is only temporary. Although the overall costs of maintaining geoscience data and collections are low compared with those of reacquisition, the amount of money a single repository requires in a short time to alleviate the space problem can be prohibitive. Moreover, even repository managers who are fortunate enough to be able to construct or acquire new space usually overestimate the length of time it takes to fill the expanded repository—and the cycle begins again (see Figure 2-3).

The problem of limited space is not unique to the United States. A survey by the Canadian Society of Petroleum Geologists (CSPG, 2001) revealed that, of 14 large companies (those with more than 200 employees), 7 had destroyed core or cuttings at least in part because of space limitations (other factors included that the materials were deemed outdated or dilapidated, or the storage cost was untenable).

ADDITIONAL SOURCES OF LOSS OF GEOSCIENCE DATA AND COLLECTIONS

Although lack of space may be the main source of loss, threats to preservation of geoscience data and collections derive from many directions. Contributory to the potential destruction and loss of the nation's geoscience legacy are: corporate mergers and restructuring, consolidation within government agencies and subsequent modifications to their chartered responsibilities, university and museum funding pressure, and retirement or reassignment of personnel, among many other examples (see Table 2-4).

Industry

Downsizing, consolidating, and public attitude toward exploration for and production of domestic resources have changed the basic structure and operating strategies for the petroleum and minerals industries. An increasing percentage of the exploration and operating budgets in both of these sectors is being re-directed to foreign ventures. The preservation of geological, *geophysical*, and *geochemical* data that have been collected by these companies is jeopardized by the cost to house and maintain domestic archives. Companies often consolidate to operate more efficiently at reduced cost. To do so usually means that something must be cast off.

Petroleum Industry

The largest domestic geoscience data and collections repositories are held by the major oil companies (ExxonMobil, Shell, BP Amoco, ChevronTexaco, and ConocoPhilips). These and other major oil companies collectively have spent billions of dollars obtaining, preserving, and curating large collections. To a great degree, the continued success of these large corporations rests on their recognition of the value and repeated use of their collections. These collections are commonly housed near research facilities where the raw material

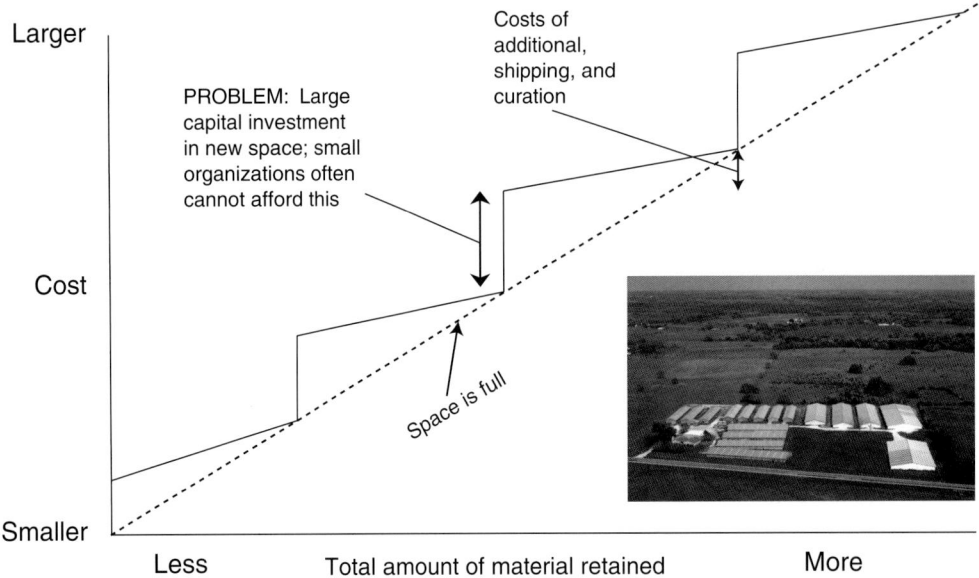

FIGURE 2-3 Cost of archiving geoscience data and collections versus total amount of material retained. Sharp increases in cost occur when capital expenditures are needed for new space. Inset: Aerial photograph of C&M Storage, Inc., the most recent buildings on the right-hand side of the image. Each additional building in this image would represent a vertical step in the main figure. SOURCE: American Images, Marshfield, Wisconsin.

may be easily accessed for study. As the focus of petroleum exploration turns international and to the deep offshore, however, interest in domestic onshore and shallow-offshore collections will wane and these collections may be lost.

Over a period of years and multiple transactions, consolidated and downsized companies can lose data and collections simply through inability to track them (see Sidebar 2-3a). The trail of merged companies can be extensive. For example, EEX Corporation represents the amalgamation of 22 entities, and a single data administrator is charged with organizing data records of all 22 companies (Michael Padgett, EEX Corporation, personal communication, 2001). In another common situation, a company may decide to move its operations. At this point, data and collections often either are divested or placed in long-term storage. The risk of loss is compounded by the fact that most current repositories already are nearly full (Tables 2-3a,b) and may be unable to accept donations. A repository that can accept the contribution may be distant, which often means prohibitive crating and shipping expenses (see chapter 5), and the risk of damage during handling and transportation. Moreover, most repositories that can accept data and collections typically limit donations to those within their region or state. Finally, company staff who are most familiar with a collection and represent much of the *institutional memory* may be lost through downsizing or attrition.

Of the threats to preservation of geoscience data and collections outlined in Table 2-4, a number are illustrated in Sidebar 2-3b. The decision makers in this situation were not sufficiently informed of the geologic and economic significance of the cores, and time and money were therefore lacking for an adequate solution to the problem. Ultimately, the result was costly redrilling of wells in the area represented by the discarded cores.

Federal tax laws and regulations also are considerations in the donation of geoscience data and collections. Guidance

TABLE 2-4 Threats to Geoscience Data and Collections (in alphabetical order)

Changing interests of some companies away from domestic exploration
Company mergers and internal management priority changes within companies
Decision makers not properly informed about geologic relevance
Deterioration of materials or metadata over time
Difficulty enforcing submission of required data and information on new materials
Inadequate supporting information on existing samples (i.e., bad metadata)
Lack of clear incentives to preserve samples
Lack of expertise to evaluate materials
Lack of space in existing repositories
Penchant for collecting new information versus working with existing information
Perceived legal liabilities
Perceived ownership by researchers instead of institutions
Reduction in force and other unreplaced departures
Retirement or departure of staff without capturing their knowledge
Samples pass into private collections
Technology changes and data are not converted from old, obsolete formats
Traditional archives (and libraries) are not interested in some collections

---

> **SIDEBAR 2-3**
> **Regrettable Losses**
>
> a) *North America's deepest well:* From 1972 to 1974, Lonestar Petroleum Company drilled the Bertha Rogers #1 oil well in Washita County, Oklahoma. This well was drilled to a measured depth of 31,441 feet, the deepest in North America. The core and samples changed hands over time. Lonestar was absorbed by Enserch, then by EEX Corporation, then by Lariat Petroleum Company, and finally by Newfield Petroleum Company. A dispute over warehouse fees during an office move resulted in a mix-up. The warehouse owner discarded the samples. The cost to re-collect these samples today would be between $12.3 million and $16.4 million (SOURCE: Michael Padgett, EEX Corporation, personal communication, 2002).
>
> b) *Core discarded into Long Beach Harbor:* In 1978 the Long Beach Harbor Department decided to locate a bridge over some railroad tracks. In the construction path was an incinerator plant that had been re-configured and was now being used as a core facility for the Wilmington oil field. The building had to be razed and the cores had to be removed. A decision was made that the cores should be discarded to save money. Mr. Mel Wright, chief geologist for the Department of Oil Properties, led the effort to save the cores. Several cores were sent to the California Well Repository in Bakersfield, but the repository was unable to accept more because of space limitations. Because funds were not allocated to relocate the cores to a new facility, transportation costs were donated by one of the oilfield service companies. After some energetic scrambling, two shipping containers were obtained and the samples were preserved, along with a few selected core sections. The rest of the cores (122 wells, at least 8 of which were cored almost continuously to depths of 8,000 feet) were deposited in the fill that became the Long Beach Harbor expansion. Lost were hundreds of thousands of feet of useful core.
>
> Unfortunately, the containers with the surviving cores were moved several times and in the process the cores were jostled and destroyed. A few years later it was determined that additional cores were needed from the same sites and same areas to replace the lost cores. Fewer than 10 wells were drilled at a cost of $9 million. The original 122 wells cost approximately $1 million to drill. A comparatively modest investment in preservation would have resulted in an order-of-magnitude increase in available data (fewer than 10 versus 122 wells)
>
> SOURCE: Mel Wright, City of Long Beach (retired), personal communication, 2001.

from the federal government on the donation of geoscience materials is key to successful transfer of collections from the private to the public domain (Kenneth Telchik, IRS, personal communication, 2001). Companies also may be wary of potential legal liability for donated material EPA considers hazardous. Petroleum residue, for example, if present in sufficient amounts within a rock core, renders the core hazardous at the time of disposal (Resource Conservation and Recovery Act, 42 U.S.C. § 6901 et seq. [1976]).

Small petroleum companies and individuals also have useful data that warrant saving. The owners may be willing to donate these materials, but usually without supporting funds, thus decreasing the likelihood of acceptance. Sidebar 2-4 includes a notable exception to this generality.

Minerals Industry

Unlike the situation in many other countries (see Sidebar 2-5), the United States has no requirements governing the disposition of geoscience records, reports, or collections from public lands obtained during the course of mineral exploration or mining. Many of these exploratory activities produce large volumes of geoscience data and collections (see Sidebar 2-6) that typically are not publicly available. Although drill samples may be retained by a company for a

> **SIDEBAR 2-4**
> **Dibblee Foundation: Ensuring Knowledge Transfer**
>
> In California, a group of geologists established the Thomas Wilson Dibblee, Jr. Geological Foundation. The foundation's goals are to preserve and publish, and thus make available, the unpublished geological maps of Tom Dibblee, the preeminent field geologist in California history. The foundation has been active for 18 years and has published 76 maps to date. These maps are widely used in California by various groups, including the USGS, the U.S. Forest Service, municipalities, counties, and consulting geologists. Another 500 maps await attention. As 2001 ended, the Dibblee Foundation entered into negotiations with the Santa Barbara Museum of Natural History to take over the effort. Part of the donation of maps will include the necessary funding to continue capturing, preserving, and distributing the vast knowledge accrued by just one person over a very large area.
>
> Committee Conclusions of Best Practices: (1) community (user-driven) involvement; (2) financial contribution to support preservation and publication efforts.

> **SIDEBAR 2-5**
> **Australian and Canadian Assessment Reporting Requirements:**
> **A Contrast to Those in the United States**
>
> In mineral exploration it is commonly the third- or fourth-generation explorer on the same piece of ground that ultimately makes the discovery. The reporting systems in place in Canada and Australia make the discovery process more efficient than in the United States.
>
> Current regulations for mining claims on public lands administered by the Bureau of Land Management or the U.S. Forest Service do not require the filing of geologic information collected on these lands as part of the annual assessment requirement. As a result, information remains the property of the exploration group and may be destroyed or lost once the claims become invalid and the lands become open to the public again. In contrast, mandatory reporting of information serves to supplement existing datasets when mineral exploration is carried out on public crown lands in Australia or Canada. For example, the Australian Minerals Act of 1978[a] has as its goal "to improve understanding of property." Reports on all wells must be sent to the state government in digital format. These are held as proprietary (i.e., they remain closed) as long as the property is in private control. This process results in a growing archive of geologic information that is released from strict confidentiality whenever control of a property returns to the public domain. The full historic database becomes available for inclusion in federally or state-sponsored studies as well as to parties interested in conducting further geologic exploration on these lands.
>
> Australia and Canada now have the ability (and in the case of Australia, the requirement) for users to submit an annual work assessment report digitally. This report must include information such as geologic, geochemical, geophysical, or sample location maps, copies of assay and drilling reports, location and type of drill holes, drilling angle, logs of rock type for all drill holes, and results of any downhole surveys. The required content of the report varies among provinces in Canada (Don Birak, AngloGold North America Inc., personal communication, 2001).
>
> Some perceived disadvantages of the Australian and Canadian requirements for filing, archiving, and accessing geologic information are the burden of additional reporting, and concerns over loss of confidentiality (Don Birak, AngloGold North America Inc., personal communication, 2001). However, advantages to exploration companies include full access to the information on public lands, which leads to more timely data gathering, reduction of duplicative effort, and reduced costs. On the government level, full data disclosure of activities and findings submitted in a standard format results in ease and efficiency in data handling and updating files. For example, data gathered from other investigations, such as airborne geophysical data, can be integrated readily into preexisting map compilations to produce real-time updates. The fully compiled dataset can be re-processed as technology advances to produce better quality scientific products, and, ultimately better estimates of publicly held resources and their value.
>
> Two factors aid successful implementation of the Australian and Canadian reporting systems described above. In addition to public (crown) ownership of mineral rights in these countries, the second contributing factor is the smaller scale of activity and generation of new data compared with that of the United States.
>
> ---
>
> [a]See Australia Department of Industry, Tourism, and Resources, 2001.

short period, they commonly are discarded unless they are offered to and accepted by a university or other group (Don Birak, AngloGold North America Inc., personal communication, 2001). Most of this activity is on western public lands administered by the Bureau of Land Management (BLM) or the U.S. Forest Service (USFS). As a result, much of the geologic information for large expanses of public *terrain* in the western United States either remains in the files of mining companies and geologic consultants or has been discarded at the completion of a project. As mining becomes a smaller portion of the domestic economy and many U.S. mining companies consolidate, the transfer of geologic archives from one company to its successor can be another source of danger for data and collections. If data do not pertain to a key asset, they often are discarded.

One of the few accessible, privately owned repositories of minerals-related geologic data is the Anaconda Collection, which is preserved at the University of Wyoming (maps and reports) and at Montana Tech (rocks, cores, and samples) (see Sidebar 3-7). However, only someone who has long-time intimacy with domestic mining could follow the anastomosing trail of the files of Magma Copper (files now with an Australian company, BHP-Billiton), Cyprus Minerals (now owned by Phelps Dodge Corporation), Amax Mining (files now with Kinross Gold Corporation and Phelps Dodge Corporation), New Jersey Zinc (files owned by a private in-

> **SIDEBAR 2-6**
> **How Much Core and Cuttings Does the Average Minerals Exploration Project Produce?**
>
> The following statistics give a sense of the volumes of material generated during minerals exploration. A small, single failed minerals exploration drilling program may generate 5,000 feet of continuous core from shallow holes at costs on the order of $350,000 to $500,000.[a] A similar program relying on rotary drilling would produce about 20,000 feet of chip samples spread over 30 to 40 holes at similar costs. A successful exploration project that evolves into a new mine commonly requires several hundred drill holes to adequately assess its economic potential. The drilling-related costs alone for a medium-sized project are on the order of $15 million to $30 million. Such a project would generate about 250,000 to 500,000 feet of chips and 50,000 to 200,000 feet of continuous core. A large project could easily double these numbers. However, subsurface exploration and development do not stop even after a mine opens. Ongoing exploration and development at an operating mine will incur annual, ongoing drilling costs of $1 million to $5 million to search for sustaining ore bodies. With success, drilling programs again expand to levels of tens of millions of dollars and many tens of thousands of feet of recovered drilling materials.
>
> [a]The costs are for the core-drilling component only. Typical overall project costs range from $1 million to $2 million.

dividual in the United States), Inspiration Copper (location of files not known), St. Joe Minerals (files now with Barrick Gold Corporation, and Doe Run Mining), and many others, as well as the now defunct minerals divisions of many oil companies. (For example, Standard Oil files are now with Rio Tinto Ltd.; Chevron files are in a number of hands, the two most important being AngloAmerican and Ivernia West Inc.) Consolidation and preservation of these data are imperative because they provide critical insight to the long-term supply of many strategic mineral commodities.

The Academic Sector

In the academic sector, scientific data and materials that underlie research reports likely will be lost over time. Scientists who assembled the data and collections usually retain possession of them, irrespective of the source of financial support for investigations (public or private). Many researchers harbor an intense sense of ownership of these materials. Furthermore, these data and collections sometimes are inadequately documented, which becomes a particularly visible problem after retirement or some other form of departure. Useful information is lost because of poor documentation, poor storage, and poor accountability. Another threat is the lack of allocation of funding for core and sample storage and maintenance within departmental operating budgets (Wayne Ahr, Texas A&M University, personal communication, 2001).

Some geoscience collections in academia (and elsewhere) are referred to as orphaned or endangered. ***Orphaned collections*** are collections of scientific value that are no longer wanted by the institution or individual that houses them, and the institution or individual, either publicly or de facto, has renounced its responsibility to care for the collection (Allmon and Lane, 2000; see Sidebar 2-7, for example). ***Endangered collections*** are those that lack curatorial support at the moment or are in imminent danger of permanently losing curatorial support. Such collections are particularly common at universities and colleges, especially when a faculty member who may have built or cared for the collection retires or leaves. Such collections either are discarded or adopted by a museum or, more rarely, another university. Museum collections rarely become orphans, unless a museum closes or changes its mission or scope of collections.

No national protocols have been established to find permanent homes for collections that become orphaned. The staff left behind may have little scientific expertise or interest and may make little or no attempt to find a permanent home. Such collections commonly languish and deteriorate until finally they are discarded (see Sidebar 2-8). Universities, in particular, have limited space and increasingly tight budgets. New faculty members who replace retirees need space for their own research. The death of a faculty member may exacerbate the problem if the person was the sole advocate for preservation of the collections. Orphaned collections that are rescued and adopted almost never come with funding from their institution of origin, which may make it difficult or impossible for a potential adopting institution to take on a collection, especially a large one. For many years, the National Science Foundation (NSF) funded other institutions' adoption of orphaned collections. By the mid-1990s, however, NSF began to express strong reservations about providing funds to support the transfer of collections without careful justification for the merit of preserving a particular orphaned collection. Since the 1990s, several systems have been established to identify endangered and orphaned collections, mostly on the Internet (e.g., the Interactive Collections Availability List [ICAL]; see UCMP, 2002a) but the success of these systems has not been quantified.

The Government Sector

The scope and priorities of government agencies that deal with geoscience data and collections have changed with time, particularly at the federal level. For instance,

> **SIDEBAR 2-7**
> **Extracts from an E-Mail Notice Sent by Killam Associates of Millburn, New Jersey, to the Paleontological Research Institution, Ithaca, New York (25 October, 2001)**
>
> [What happens to the materials discussed below if they are not accepted by anyone?]
>
> "....The [New York City Department of Environmental Protection] NYCDEP has identified surplus rock core from its Water Tunnel #3 Project. The NYCDEP is offering to donate this core to local educational institutions, museums, and Geological Surveys to fuel geological research and foster educational programs.
>
> This core was collected by the City of New York to aid in the planning and design of its drinking water supply system. Currently the NYCDEP is in possession of nearly 80 surplus borings, each of which contains between approximately 400 to 750 feet of NX-sized (about 2.2" diameter) core. The borings are from various locations throughout Brooklyn, Queens and the Bronx, New York.
>
> Sometime within the next 6 months the city expects to send out a letter to parties potentially interested in obtaining this core for their own use. Logs and other information will likely be provided at that time. If you would like to be put on our mailing list of potential core recipients, please reply with the name of a contact person, a mailing address, and a contact phone number."

when federal agencies were consolidated several years ago, many of the functions of the U.S. Bureau of Mines were folded into the U.S. Geological Survey (USGS). With the evolving mission of the USGS, and the near-simultaneous reduction in force (Figure 2-4), their ability to focus staffing efforts on geoscience data and collections management has been hampered severely. Sidebar 2-9 illustrates the influence of the current interpretation of the Organic Act on the flow of geoscience data and collections from the USGS to the Smithsonian Institution. In a parallel trend to that of USGS staffing, the Smithsonian's Collection Management program in NMNH has been unable to replace staff who retired or resigned in the last 10 years, which has resulted in a significant decline in the rate of cataloging (committee survey response, 2001).

The U.S. Army Corps of Engineers (USACE) collects geoscience data at each of its project sites. These projects involve construction of dams, levees, or other engineering structures. In total, more than 500 projects involving geologic investigations have been carried out around the country, and data currently are being collected at more than 35 sites. These data are housed at district offices, of which there are 40 across the country. Under USACE regulation (Engineer Regulation [ER] 1110-1-1803), rock cores and other geologic information must be retained for 5 years, or longer if litigation is ongoing. In practice, most cores are collected in the investigation phase of a project and are kept through construction and any litigation phases. This is typically a period of 10 years. Because there are no requirements to keep core beyond any litigation phase, there are no regulations to prevent core from being discarded. Of the 15 responses the committee received from USACE district offices, only 4 indicated success in giving away core to other groups (universities and state geological surveys). None of the four districts with the largest holdings of core had success in donating materials. At least 75 percent of core collected before 1985 has already been discarded (Michael Klosterman, USACE, personal communication, 2001). The period before 1985 coincides with that of USACE's greatest project activity. The problem is exacerbated by lack of a central USACE database of holdings to track or allow searches of materials. Instead, information is held in paper files, microfiche, or in a variety of types of computer software, primarily at individual regional offices. Financial support for publishing geologic results has not been forthcoming within USACE, hence much of the wealth of geologic information USACE gathered has never been shared with the broader geologic community. Given that USACE geoscience data and collections likely have direct relevance to engineering issues and societal needs (e.g., dams, levees, roadways) the loss of this information is particularly troubling.

The committee concludes that no agency in the federal government is charged with keeping all national collections of scientific value, nor should it be. However, the committee also concludes that most agencies in the federal government that keep collections of scientific value are inadequately supported to do so, or even to evaluate the collections using criteria such as those outlined below.

INACCESSIBLE GEOSCIENCE DATA AND COLLECTIONS

Even if geoscience data are not permanently lost or destroyed, they may still be inaccessible to the public. Three broad categories of data and collections that currently are inaccessible to the public are: those temporarily lost, those held by companies or individuals and considered proprietary, and those in the public domain, but neither cataloged nor curated.

Data and collections in the public domain that are not truly accessible include information that has been stored

SIDEBAR 2-8
Examples of "Lost" Fossils

Reliable details are elusive for fossil collections that institutions have discarded. Anecdotal accounts, however, are so numerous that it is reasonable to conclude that these losses are not uncommon. On further investigation, some prove to be apocryphal; others can be neither verified nor refuted. The common theme from the three examples in this sidebar is that the fossils evidently underwent no formal ***deaccession*** process.

- In the late 1800s, a large specimen of the giant ground sloth *Megalonyx jeffersoni* was on display at the Indiana University museum (see image below). Although a fire destroyed the museum, the specimen was complete and intact at the turn of the century. By 1901, there was not a single museum room at Indiana University. Instead, several departments each had their own small museum. Sometime between 1937 and 1947 the *Megalonyx* was dismantled and either lost or discarded, except for 5 of the 60 or so bones. Why did this occur? As is typical with lost or discarded specimens, space for display may have become a problem. In this case, the space problem likely resulted from return of World War II veterans to America's colleges and universities. Another possibility is that no one who cared greatly about the specimen was around to defend it after a new geology department chair arrived at Indiana University in 1945. Reportedly a dump truck backed up to the department building, and students and faculty tossed unwanted specimens out a second story window into it. Whatever happened to the specimen, the story is far less atypical than one might imagine, and the circumstances surrounding its demise still hold for geoscience data and collections at risk today (Lane, 2000, p. 23–29).
- When the Boston Society of Natural History moved into a new building in the 1950s, eventually becoming the Boston Museum of Science, they had large collections of dinosaur tracks. Some were transferred to the Museum of Comparative Zoology at Harvard and the American Museum of Natural History. The rest were given or traded to a commercial collector in South Hadley, Massachusetts. At least one large slab went into a landfill (Emma Rainforth, Columbia University, personal communication, 2001).
- A post-doctoral research fellow at the Smithsonian amassed a substantial collection of fossils. Most were on slabs, hard ground, and rock pavement, but all were well located and identified. They were large and inconvenient and did not fit into the Smithsonian's drawers. Consequently, the museum apparently wanted to discard the slabs. A researcher who accidentally stumbled upon this situation arranged for the material to be donated to Wooster College, Ohio. Without the chance discovery of the threat and the researcher's intervention, the slabs would have been lost because they were inconvenient to store and had no champion from within the museum (Tim Palmer, University of Wales, personal communication, 2001).

Megalonyx jeffersoni in the Indiana University Museum in the late 1800s. Loss of large, complete specimens such as this is especially tragic because of their rarity and scientific value. SOURCE: Indiana University Archives, Bloomington.

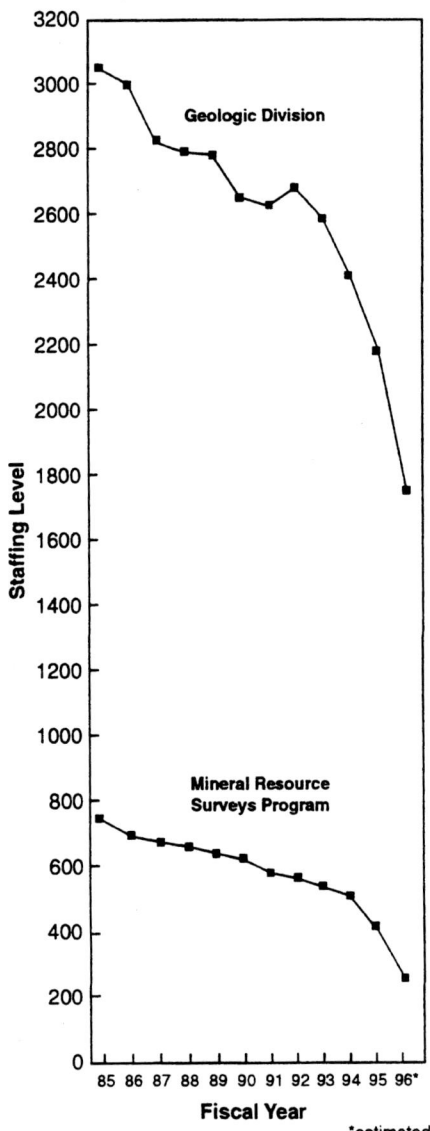

FIGURE 2-4 Staffing-level trends in the USGS's Geologic Division and Mineral Resource Surveys Program (MRSP) from 1985 to 1996. From 1994 to 1996, the Geologic Division staff dropped by about 27 percent and the MRSP staff fell by 49 percent. Current staffing levels for both groups (the MRSP is now called the Mineral Resources Program) remain close to 1996 levels (Linda Gundersen and Kathleen Johnson, USGS, personal communication, 2002). SOURCE: Eaton, 1996; NRC, 1996b, p. 7; unpublished data provided by the USGS.

improperly, is not cataloged, lacks documentation, or is not well curated. If the material cannot be found, it is useless (see Sidebar 2-10).

PRIORITIES FOR PRESERVATION OF GEOSCIENCE DATA AND COLLECTIONS

The preceding sections of this chapter have outlined many factors that have led to the indiscriminate loss of geoscience data and collections with even more at risk of being lost. In such situations, a critical decision for those possessing the data becomes whether to retain or discard the data. Those who may be offered at-risk data face a similar decision: they must decide whether to accept or refuse the data. Space and cost commonly dictate the outcome of these management decisions. A well-rounded decision, however, can be made only if priorities are set for what to preserve.

In the course of setting priorities for accession and deaccession of geoscience data and collections, it became apparent to the committee that the broad range of data and collections precluded assignment of evaluative criteria across the board. Table 2-5 illustrates this point. Quality and completeness are less an issue for cuttings, which typically are less complete than cores. Moreover, cuttings tend to mix as they make their way up a drill hole. In contrast, it is the quality and completeness aspect of sediment and ice cores that makes them unique and powerful storehouses of important paleoclimatic information (among other types of information they record). For all other collections listed, the range of acceptable quality and completeness is variable and best left to those who know the most about what is acceptable at various levels (legal versus research versus teaching, in approximate decreasing quality control order). Accuracy is yet another metric that may not be a factor for some types of geoscience data and collections. For instance, maps, notes, and other unpublished materials may be highly inaccurate, but their historical context (if well documented) could be very valuable in understanding how someone was led astray. For geophysical information, the accuracy could be very poor, but some valid information can be extracted mathematically from even highly inaccurate data.

Replication of geoscience data and collections (i.e., multiple samples of the same or nearly the same item or items) probably is the most contentious of the criteria listed in Table 2-5. For very different reasons, having replicates of some geoscience data and collections actually can be a positive factor instead of a negative factor in their retention. For instance, replicate or nearly replicate information for engineering collections (i.e., multiple drill stem tests from the same well or from nearby wells) can be extremely useful in assembling the history of reservoir development and exploitation. Multiple specimens of the same fossil taxon allow evaluation of population information such as variability, which is assessed as a factor in whether a similar fossil is a new taxon or simply within the range of shapes that one might find in another taxon. Multiple fossil specimens also provide information about abundances of taxa, which are of fundamental importance in population and extinction dynamics. Mining cores can differ from most other cores in that, if a mine is opened and part or all of a deposit mined, the mine cores may be the only remaining record of the

---

[3] Furthermore, such cores have continuing value by providing evidence of unique geologic conditions that combined to form a mineral deposit.

SIDEBAR 2-9
The National Museum of Natural History and the U.S. Geological Survey

The Smithsonian Institution has special relevance to the issue of geoscience data and collections preservation for at least three reasons: it houses the largest geoscience collection in the world; it serves as the national museum for the United States; and it has a long statutory connection with the USGS with regard to collections. (The Smithsonian was founded in 1846. The U.S. National Museum [USNM] was founded within the Smithsonian in the 1850s and ceased to exist in the 1970s, becoming the National Museum of Natural History [NMNH] and the National Museum of American History. The U.S. National Museum continues today only in the anachronistic acronym USNM on catalogue numbers in the NMNH collections.)

The Organic Act of 1879, which established the USGS, states that all collections of fossils of the U.S. government, including those of the USGS, "when no longer needed for investigations in progress, shall be deposited in the National Museum." For most of its history, the USGS paleontological staff and collections have been closely connected to the NMNH. This relationship changed in 1995, however, with a reduction-in-force at the USGS and the USGS's decision to dispose of much of its fossil collections. In 1996, the NMNH and the USGS signed a memorandum of understanding (MOU) describing how the NMNH would dispose of the USGS collections. Under the MOU, the NMNH would take what it wanted and the remainder would be made available to other institutions, to be selected according to a set of criteria. Under this arrangement, the USGS–Menlo Park collections were transferred to the University of California Museum of Paleontology. Most of the USGS–Reston collections were transferred to the Virginia Museum of Natural History. Having now expired, the MOU is currently being renegotiated (Ross Simons, Smithsonian Institution, personal communication, 2002).

Smithsonian staff currently (in 2001) interpret the Organic Act legislation as constituting a right of first refusal for USGS specimens. The Smithsonian has neither the space nor the scientific interest to accept all of the specimens USGS or anyone else might offer. The NMNH's Department of Paleobiology reports that it has limited space for additional collections growth. Therefore, in practice, the Smithsonian continues to add to its collections, but to be highly selective in doing so. The NMNH sees itself as the keeper of the nation's treasures, not the nation's collections. The NMNH does not see itself as a repository of last resort for all orphaned collections or as the ultimate repository for all of the national collections generated or formerly housed by other federal agencies. In other words, it may accept, but is not obligated to take, collections from the USGS.

The USGS follows Department of the Interior policies on museum property and its Museum Property Program requires accountability for all historical and museum collections under the bureau's control. Research collections do not fall under this program, and USGS currently is developing a policy for managing these working collections (Allan Montgomery, USGS, personal communication, 2002). In practice, the USGS appears to have had limited interest in maintaining specimen collections for the long term. For example, the USGS employs only three staff members (down from eight in 1994) dedicated to preservation of geoscience data and collections at its Lakewood repository (Sidebar 3-2), and the storage space at that repository was reduced by 40 percent in 1995. Furthermore, volunteers curate the USGS's irreplaceable, nationally ranked (and federally owned) paleontological collection. Lastly, as noted above, the USGS has given away two-thirds of its paleontological collections over the past 10 years.

The committee visited the Smithsonian Institution in April 2001, and the USGS Lakewood facility in June 2001.

SOURCE: Questionnaire responses and input during site visits.

---

mined material.[3] Consequently, multiple mining cores are less an issue than multiple rock cores from non-mined deposits. Finally, sediment and ice cores have an inherent fragility and sensitivity to storage conditions that make replicates (especially if kept in a separate facility or separate part of the same facility) a wise insurance policy against intellectual loss. In addition, seemingly identical sediment and ice cores from geographically separate areas are critical in assessing the geographic range and potential global impact of various climatic and climate-related events on Earth.

Upon inspection of Table 2-5, one might wonder about seemingly obvious criteria that are missing. One such criterion is use as a factor in assessing the importance of a single specimen, a group of holdings, or an entire collection. Use is an especially poor criterion for assessing priority for two reasons. First, use commonly relates to how well known (or

SIDEBAR 2-10
Examples of Inaccessible Geoscience Data and Collections

1) The paleontological collection at USGS Denver Federal Center is probably the largest such collection in the United States for which there is no funding for curation. It is also one of the largest with no standardized, computerized catalog. Knowledge about the collection resides with only a few people, many of whom are retired. As large and scientifically important as the USGS fossil collection is, there is no budget for collections management (committee survey response, 2001). Staff paleontologists have direct responsibility for curating their own collections (each cataloging specimens in their own style), yet official allocated time for curation is zero. The result is a variety of catalogs in handwritten ledgers, typed index cards, or computer database systems with no standardized format or medium of storage. Individual collections are commonly accessible only when the investigator is present. When a scientist retires or leaves, much of the institutional memory about the collection also departs. (The committee visited the paleontological collection in June 2001.)

2) An independent oil company, HS Resources, acquired Amoco's interests in an oil field in 1997. By June 2001, after being stored outdoors for 2 years, the cores were on unorganized pallets in a warehouse with random equipment laid on top of them. HS Resources merged with Kerr McGee in September 2001. The cores were still in the same location in February 2002 (John Ladd, Kerr McGee Rocky Mountain Corporation, personal communication, 2002).

3) DOE cores stored at Oak Ridge National Laboratory (ORNL) in Tennessee are stacked outside buildings in the open air and are overgrown with weeds (see photograph below). If they are not curated soon, these cores will be useless. Even if the rock should survive, the boxes and annotations on the samples will be lost, thus rendering them nearly valueless. The cores in the photograph in this sidebar were obtained by the Tennessee Valley Authority on the Clinch River Breeder Reactor site. Cores at DOE's Hanford, Washington, site also are exposed to the elements, although they are not maintained as poorly as those at Oak Ridge. In addition, the operating contractor in charge of ORNL, University of Tennessee–Battelle, has indicated that it may dispose of all but a few thousand feet of the 35,000 feet of rock core for the Oak Ridge Reservation—samples that indicate the fractured rock characteristics and basic subsurface geology for the ORNL site. The replacement cost for these cores is estimated at $5 million to $10 million (Richard Ketelle, Bechtel Jacobs Company LLC, personal communication, 2002).

Cores stored outside at Oakridge National Laboratory, Tennessee, in Spring 2001. These cores are from the Tennessee Valley Authority's Clinch River Breeder Reactor site. Rescue of cores in this state of degradation is unlikely given the probable loss of documentation associated with them. SOURCE: Richard Pawlowicz, Bechtel National, Inc., San Diego, California.

TABLE 2-5 Criteria for Determining Which Geoscience Data and Collections to Preserve

Criteria	Well Documented[d]	Irreplaceable[e]	Potential Applications[f]	Accurate	Quality/ Completeness	Non-Replicative
Collections:						
Cuttings	X	x	x	X	_	X
Engineering[a]	X	x	x	X	x	_
Fossils	X	x	x	X	x	_
Geophysical[b]	X	x	x	_	x	X
Maps/Notes[c]	X	x	x	_	x	X
Mining Cores	X	x	x	X	x	_
Other Rock Cores	X	x	x	X	x	X
Sediment & Ice Cores	X	x	x	X	X	_

X = present or necessary for preservation (i.e., absence = candidate for deaccession).
x = may be present and may be a factor for preservation (i.e., absence may not be a factor for deaccession).
_ = not present and not necessary for preservation (i.e., absence is not a factor in deaccession).
Criteria are arranged from left to right in approximately decreasing order of importance (but see text for further explanation and elaboration).
Collections are arranged alphabetically.

[a]Includes drill stem tests, completion records, site reports, and other engineering data/reports on CD, computer disk, fiche, paper, tape, or some other quasi-stable medium.
[b]Includes seismic data, down-hole geophysical data, fly-over geophysical data, and other geophysical data on CD, computer disk, fiche, paper, tape, or some other quasi-stable medium.
[c]Includes unpublished materials on CD, computer disk, fiche, paper, tape, or some other quasi-stable medium, whether or not they were used in the production of published products.
[d]All collections must be well documented before any other assessment of their utility and future can be done. Indeed, whether or not a rock, fossil, core, or other item is replaceable or not is completely unknown in the absence adequate documentation to assess uniqueness. That said, if part of a collection is not replaceable, but only documented well enough to know that it is unique, it probably should be kept anyway. Documentation includes, but is not limited to, information about age, location, depth, collector or author, date acquired, and associated materials.
[e]Impossible or highly unlikely to collect a similar sample (e.g., a mine core from a completely mined-out locality; a sample from a politically inaccessible part of the world; a sample requiring great time and effort to recollect such as a deep ice core from Antarctica or Greenland).
[f]This category in particular should be weighed judiciously by a science advisory board comprised of members of the user community.

not) a collection is. A critical collection, key fossil, or pivotal core may be completely unused if its whereabouts is unknown to most people. This most often occurs because of inadequate metadata (data about the collection). Clearly, in these instances, if the collection were known, it would be used. Consequently, its lack of use is an inappropriate measure of its importance or future relevance (if appropriate metadata are provided).

Use statistics are inappropriate for a second reason; immediate use is not necessarily an indicator of future use, even if the metadata are well known and well established (see chapter 1 for examples of unanticipated use). Future use is difficult to predict, but almost always hinges on the otherwise assessable criterion of documentation. Poorly documented geoscience data and collections almost never have any future.

Also not present in the criteria listed in Table 2-5 is cost. The committee specifically avoided the issue of cost in determining which geoscience data and collections to discard and which to keep because this is best determined at a local level.

Table 2-6 summarizes general guidelines for assessing donation and reception priorities for donors and recipients of geoscience data and collections. Table 2-5 and Table 2-6 should be used in concert with each other. They also should be used with caution. It was neither the committee's desire to be overly prescriptive or limiting about setting priorities for accepting geoscience data and collections, nor was it the committee's intent that these criteria be applied without consideration and input from user communities. For this reason, *the committee concludes that close, meaningful involvement of external science advisory boards, which includes membership of an expert able to assess metadata issues and other issues of discovery and accessibility, is vital.* Sidebar 2-11 illustrates the role of such a board in advising the managers of the National Ice Core Laboratory in Lakewood, Colorado. The science advisory structure for the Ocean Drilling Program is illustrated in Figure 4-1.

Science advisory boards are in the best position to give realistic recommendations (as opposed to the unrealistic recommendation of keep everything) about what to keep against the backdrop of what might be needed in the future. Because of the complexity of such decisions, they should never be left to any single person. Broad, community-based input using community-driven criteria is the best approach for assessing which geoscience data and collections merit retention and which should be discarded.

SIDEBAR 2-11
Managing Ice Cores at the National Ice Core Laboratory

The National Ice Core Laboratory (NICL), at the Denver Federal Center in Lakewood, Colorado, manages ice cores collected and used primarily by NSF- and USGS-funded researchers, and other government personnel. A web-based catalog (www.nicl.usgs.gov) enables potential users to determine current holdings. Through an outreach program, NICL introduces people of all ages to ice-core science.

With holdings of 15,700 meters (51,509 feet) of ice core (see photograph in sidebar) at $-36°$ Celsius, NICL is currently at 90 percent capacity. Implementation of a staged plan for a new, mobile racking system will increase available space from 10 to 48 percent, thus deferring space problems for several years.

NICL has operated under an inter-agency agreement between the USGS and NSF since opening in 1993. The annual budget ($477,000 in 2000) is shared equally between these partners. USGS has responsibility for facility operations, while NSF provides oversight that includes periodic performance reviews. Science management (including decision making on sample allocation and accession and deaccession protocols) is coordinated by the University of New Hampshire under a competitive contract. The director of scientific management bases his or her guidance on advice from an Ice Core Working Group (ICWG). The ICWG is a group of 11 experts from universities and the USGS who actively work on ice cores and/or use data derived from ice cores.

Accession and deaccession protocols are promulgated by NSF using a subgroup of the ICWG as an advisory committee. This removes NICL from potential conflicts of interest on such matters. To be accepted by NICL, ice cores must arrive with site information and logging information for each meter section of core. This information should be in digital form. For smaller-diameter (4-inch) cores of opportunity, a removal date must be established upon accession. Because of the necessary high levels of coordination for deep drilling projects in Greenland and Antarctica, NICL has advance notice of incoming large-diameter (5.2-inch) cores and can plan accordingly. Currently approximately 1,000 meters per year (3,281 feet per year) of new core are collected in a typical drilling season. This plan may include deaccession of older core, a process overseen by the ICWG. Criteria for deaccession, each assessed on a scale of 1 to 5 by scientists, are age, continuity, volume, robust dating, published information, number of requests, core quality, duplication, drilling method, specific utility, uniqueness, and site accessibility. Deaccessioned ice offers a testing ground for new analytical methodologies such as extraction of CO_2 from air bubbles (Sidebar 1-7). Deaccesioned cores are advertised through e-mail and print to a broad cross-section of the scientific community. In June 2001 approximately 1,588 meters (5,210 feet) of core were on the deaccession list, and another 477 meters (1,565 feet) were shipped to scientists and school outreach programs in the preceding year. Ice is not discarded until NICL needs the storage space, and only then after many others have passed on the opportunity to take the ice themselves.

Committee Conclusions of Best Practices: (1) web-based catalogue of metadata; (2) inter-agency and federal/university support; (3) community-based, science- and user-oversight committee; (4) well-documented and well-advertised deaccession protocols that result in little wasted core; (5) adequate fiscal support (as of 2001).

The committee visited NICL in June, 2001.

Inside the National Ice Core Laboratory. The ice core storage room is maintained at a temperature of $-33°F$ ($-36°C$). About three-quarters of the NICL collection is in 1-meter (3.3-foot) tubes, the other quarter is in 1.5-meter (5-foot) tubes. Each tube contains part of a single core. SOURCE: Geoffrey Hargreaves, NICL.

TABLE 2-6 Guidelines for Assessing Donation and Reception Priorities for Donors and Recipients of Geoscience Data and Collections[a]

Responsible party	Guidelines for Assessment
Donors	1. Provide digital inventory or other documentation of donated materials. 2. Document uniqueness, significance, completeness, and other known context of donated materials. 3. Ask the recipient for a written plan for curation and access in order to determine the repository's commitment to curation and access. 4. Provide financial support for transportation and curation (if possible).

Recipients	1. Assess appropriateness of donation for repository mission and/or expertise by evaluating:
	   A. uniqueness and relevance of donation vis-à-vis repository goals;
	   B. likelihood of obtaining similar material from the same place or time;
	   C. esthetic and/or preservational qualities, including completeness and significance.
	2. Provide written plan for curation and access.
	3. Assess cost (if any) to render donation useful (if it can be rendered useful).
	4. Solicit financial and/or volunteer support from the donor if required to curate the donation adequately.

Donors and recipients	1. A donation is not useful if it is undocumented. 2. A donation is a burden to a repository if it cannot be curated adequately.

[a]These guidelines should be used in conjunction with the matrix provided in Table 2-5.

3

Geoscience Data and Collections Today

INTRODUCTION

As demonstrated in chapter 2, an impressive amount of geoscience data and collections resides in repositories within the United States. The variety and types of geoscience data and collections are equally impressive. This chapter describes the major types of geoscience data and collections that are physical rather than digital, why they are collected, by whom they are collected, the nature of the data and collections themselves, and the nature of the facilities in which they currently reside.

Who Collects Geoscience Data and Collections and Where They Are Held

Geoscience data and collections are collected or held by corporations, private companies, government agencies, state geological surveys, educational institutions, public and private museums, and individuals (see Table 3-1 and Figure 3-1). While no comprehensive index of U.S. geoscience data and collections repositories exists, the American Geological Institute's National Directory of Geoscience Data Repositories (AGI, 1997) includes information on the types of data held by some repositories, along with information on the geographic area each covers.

The data and collections corporations hold are usually those acquired directly through their own activities or via purchase from other corporations. Private companies, in the form of independent repositories or data brokers, also collect and retain geoscience data and collections for sale or lease. Government agencies (state, federal, and local) collect these materials to further their scientific, economic, safety, and regulatory missions. Educational institutions and museums have similar goals, but emphasize the educational or research value of geoscience data and collections. The extent and type of geoscience data and collections acquired by these entities vary depending on their mission.

CORES AND CUTTINGS

Not all holes drilled in the Earth produce cores and not all cores are rock. Cores can consist of rock, unconsolidated sediment, or ice. Each is collected for the specific and unique information it can supply. Rock cores are long cylindrical samples of Earth's crust taken most commonly by means of a diamond core drill (for rock and ice) (Figure 3-2). Sediment cores are comparatively much shorter cylindrical samples collected most commonly by rapidly vibrating or pounding a metal tube into the sediment. Cores are collected by many different kinds of people and entities, including major and independent petroleum companies, mineral exploration companies, water resource managers, engineers, and scientists. The average oil well core is 2.75 to 4 inches in diameter and may be a few feet to a few thousand feet long (Figure 3-3a).

Drill holes (wells) are made for a variety of reasons, including: exploration and production of oil and gas; exploration for coal, metals, or other minerals; production or monitoring of groundwater; monitoring the environment; and studying rock characteristics for applied or basic research. In addition to resource assessment, examination of cores can yield essential data for study of climate change, ancient extraterrestrial impact craters, evolution of sedimentary basins, ancient and modern volcanic systems, and the deep biosphere, among many others. They also provide data essential to safely site and build nuclear power plants, dams, buildings, highways, bridges, tunnels, and other structures.

Rock cores in Earth's crust contain direct information including its **mineralogical** and **petrological** composition and structure, fluid content, fractures, fossil composition (and therefore age), and the nature of change from one rock type to another. Two particularly important features of a rock for petroleum production and water resource management are **porosity** and permeability. Porosity is a measure of the fluid storage capacity of a rock; it can be determined directly by

TABLE 3-1 Examples of Collectors of Geoscience Data and Collections, and Their Purpose

Volume of Physical Samples[a]	Public Sector		Private Sector	
	Entity	Purpose	Entity	Purpose
Larger[b]	Smithsonian Institution	Research, education	Large Petroleum Co.	Resource extraction, research
	U.S. Geological Survey	Research, education, resource evaluation		
	Large State Geol. Survey	Research, regulatory		
	Department of Energy	Research, site characterization	Large Mining Co.	Resource extraction
	U.S. Army Corps of Engineers	Site characterization		
	Ocean Drilling Program	Research, education		
	Continental Drilling Program	Research, education	Independent Oil Co.	Resource extraction
	U.S. Nuclear Regulatory Commission	Site characterization	Small Mining Co.	Resource extraction
			Private Museums	Education, research
	National Ice Core Lab	Research, education		
	University	Education, research		
	Public Museum	Education, research		
	Water Management District	Regulatory, management	Consulting Firm	Various
	Minerals Management Service	Regulatory		
	Small State Geol. Survey	Research, regulatory		
	Bureau of Land Management	Regulatory		
Smaller			Individuals	Hobby, investment

[a]Volume is estimated only from physical data retained by each group (predominantly cores, cuttings, samples) (see Table 2-1).
[b]These examples are ranked in approximate order of volume of physical geoscience collections held by each entity.

examining cores, or indirectly from examining other subsurface data. However, permeability, which is a measure of the connectivity of the pore spaces (i.e., how easily a fluid can move through the rock or sediment), can only be measured directly from examination of actual rock samples, which are recovered only in cores and cuttings from the deep subsurface. To derive particular kinds of information, cores and cuttings are subjected to a wide variety of analytical techniques, including simple visual inspection, X-raying, *CT scans*, thin sections, and permeability tests.

Ice cores and sediment cores are collected primarily because they preserve a record of past environmental change. For example, sediment cores from the ocean floor can reveal changes in ocean chemistry and, indirectly, temperature through time. Ice cores preserve ancient air bubbles, among many other useful records, allowing the determination of former levels of atmospheric carbon dioxide (CO_2) against which modern levels can be compared (see Sidebar 1-7). Our understanding of global change is grounded in the discoveries made from collecting ice and sediment cores and the historical record unlocked by those discoveries.

After cores are taken at the drill site, they usually are stored in cardboard or wooden boxes. Volume commonly is expressed as the number of boxes, or, in the case of ice and sediment cores, the number of tubes. Boxes vary considerably in size, as does the amount of core each contains. A widely used box size is approximately 3 feet long and holds three to five 3-foot lengths of rock core (9 to 15 linear feet, total) side by side within the box. Segments of ice and sediment cores are stored singly in 3-foot-long tubes. Depending on the density of the rock, sediment, or ice, each container can weigh 35 to 50 pounds. While rock cores require limited special treatment, the containers for ice and sediment must be airtight and sufficiently cold throughout transport and storage.

Not all drill holes produce core, but almost all produce cuttings. Cuttings are the chips of rock that come up the outside of the drill pipe when using any type of rotating drill bit. Cuttings are samples of the rock through which the drill bit has cut, hence their name (see Figures 3-2 and 3-3). Huge amounts of cuttings have been produced and collected from various wells drilled over the decades (see Table 2-1). Holes that produce only cuttings are cheaper and quicker to produce and collect than holes that produce cores and cuttings. This is because not all cuttings are sampled and because cuttings flow to the surface during continuous drilling, as op-

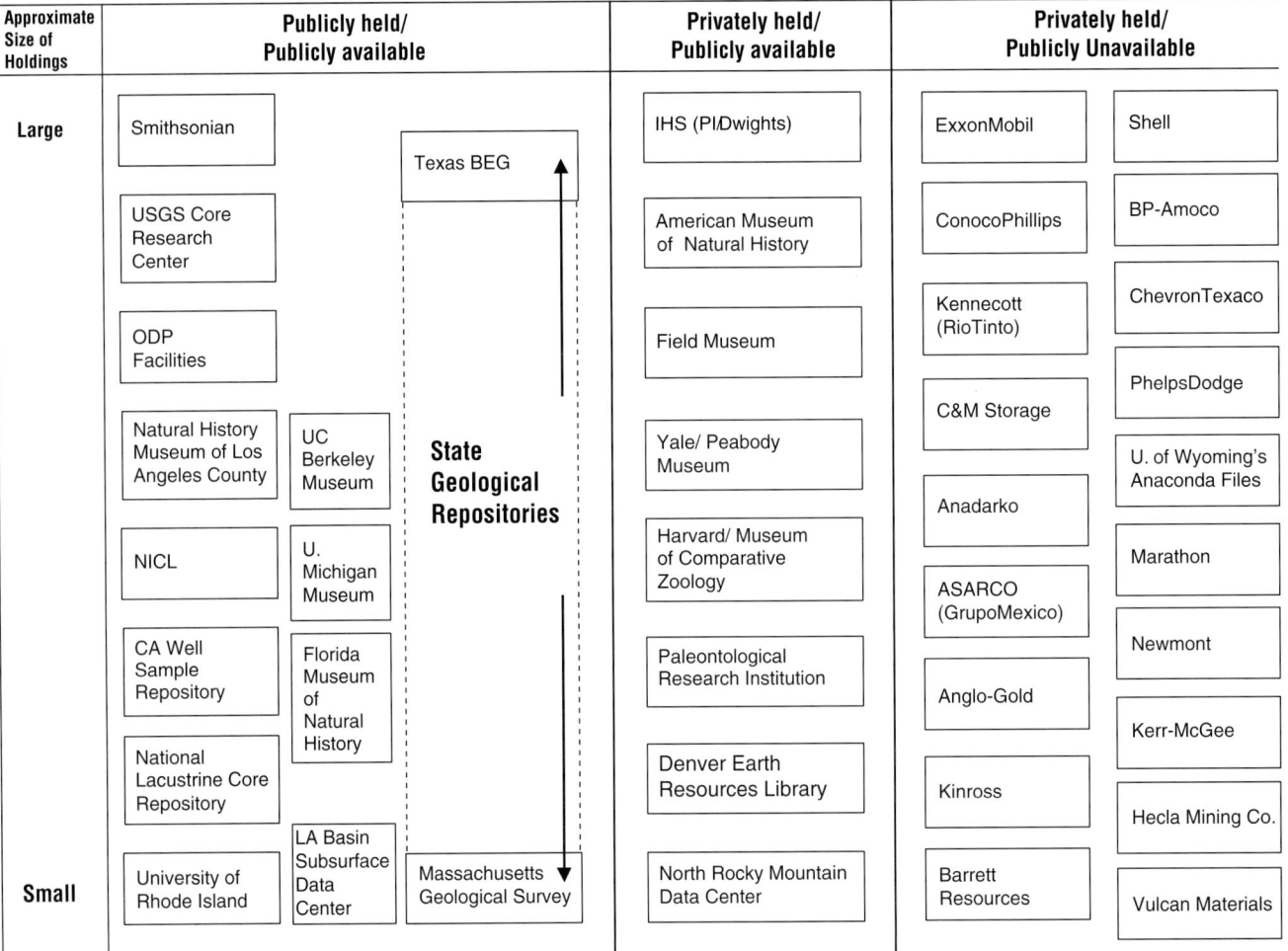

FIGURE 3-1 Examples of where geoscience data and collections are housed, arranged from large (top) to small (bottom). Archives in the private sector has two subgroups—that in which data and collections are publicly available, and that in which they are proprietary. Some proprietary holdings are maintained by public repositories, but these are uncommon and comprise only a fraction of a percentage of the overall holdings.

posed to cores, which have to be hauled to the surface between active-drilling times. When cuttings arrive at the surface, they are collected either with the surrounding drilling mud or screened out of the drilling mud and saved for later laboratory processing. Cuttings washed free of drilling mud are dried and stored in small (about 2- by 3-inch) envelopes, categorized by the depth from which they were gathered.

Although comparatively cheap and quick, cuttings still yield important information about the character and age of the rock penetrated during drilling. The use of cuttings has been somewhat limited (compared with cores), however, because of their tendency to mix with adjacent cuttings during their trip from the drill bit to the surface and because of their small size (individual cuttings typically are ¼ inch and smaller). Mixing somewhat diminishes the ability to pick precise depths of important rock units or other features of interest. The small size of individual cuttings hides recognition of some larger important features (especially fractures). Sidebar 1-6 describes new techniques being developed to extract additional information from fluid inclusions found in cuttings.

Several major non-industry projects generate significant amounts of core for basic scientific exploration of Earth's crust or ice sheets. These scientific drilling programs include the Ocean Drilling Project (ODP), Drilling, Observation, and Sampling of the Earth's Continental Crust (DOSECC, 1998), Antarctic (WAIS, 2000) and Greenland (ARCSS, 2002) ice-coring projects. The ODP and ice-coring projects serve as excellent examples of research communities that understand the importance of hard-won core and plan for adequate access and maintenance (see chapter 4).

Rock cores and cuttings are held by petroleum companies, other natural-resource companies, the USGS, state agencies, individual researchers at colleges or universities,

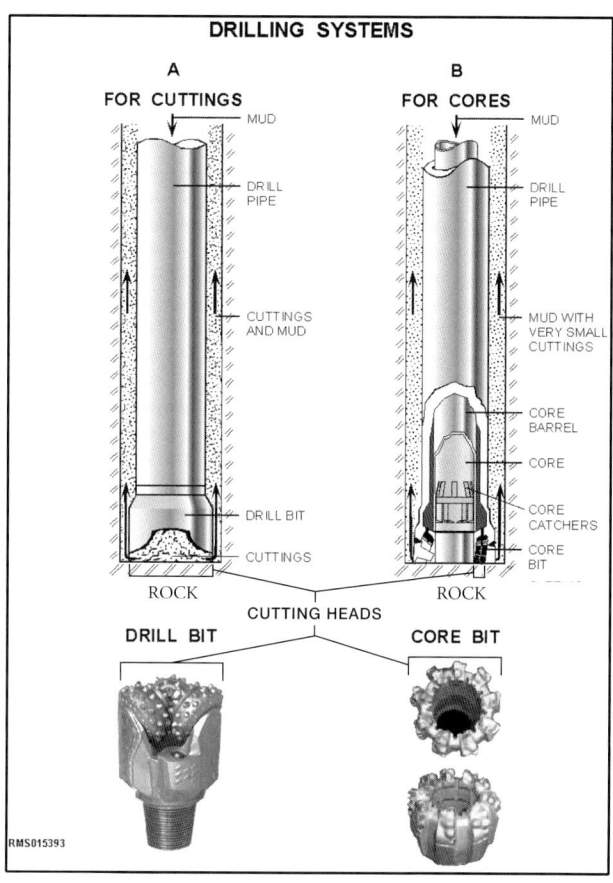

FIGURE 3-2 Coring and cutting devices. SOURCE: Baker Hughes, 2001.

FIGURE 3-3a Cores from Potter Mines, Matheson, Ontario. These cores were retrieved from a depth of 623 to 629 meters (2,044 to 2,064 feet). Each core box contains 3 meters (10 feet). SOURCE: Millstreams Mines, Ontario, Canada.

FIGURE 3-3b Cuttings. SOURCE: Baker, 1980. Petroleum Extension Service, The University of Texas at Austin.

> **SIDEBAR 3-1**
> **C&M Storage Inc.**
>
> C&M Storage Inc. is a private, for-profit company in Schulenberg, Texas, whose primary business is to facilitate the proprietary storage and retrieval of cores and cuttings collected by client petroleum companies. The facility which services 65 private companies, shuttles cores, cuttings, and other samples to and from Houston, 80 miles away, twice a week to an average of 21 clients.[a] In addition, C&M Storage provides its clients with onsite services that include inventory management, core slabbing, and geochemical sample preservation. Current storage includes more than 1 million boxes of core, cuttings, thin-section slides, paper well logs, tapes, maps, and microfiche. About 90 percent of the stored materials are cores and cuttings. C&M Storage has an annual budget of between $1 million and $2 million. Storage capacity is currently at 268,300 square feet and is expanding at a rate of about 10,000 square feet per year, sufficient to house about 125,000 new boxes of core each year.
>
> Storage facilities include a number of uninsulated, wood-framed, sheet-metal buildings constructed on leveled ground with a crushed stone flooring base. In addition, specialized storage facilities, totaling about 11,500 square feet, have been built with climate-control capability to house fragile materials and documents. As existing storage capacity is filled, additional onsite acreage remains for constructing similar buildings, each with 20,000 to 25,000 square feet of storage capacity. Facilities to lay out core for examination, with limited microscope and computer access, also exist. It is noteworthy that individual clients—not C&M Storage Inc.—make decisions on accession or deaccession of material.
>
> Committee Conclusions of Best Practices: (1) Active, supportive clientele; (2) low capital costs; (3) core, cuttings, samples owned by companies who pay for maintenance, access, service, and propriety; (4) room for growth and expansion.
>
> The committee visited C&M Storage in August, 2001.
>
> SOURCE: Robert Shafer, C&M Storage Inc., personal communication, 2001.
>
> ---
>
> [a] 61 of the 65 clients are located in Houston.
>
> C&M Storage Inc. from the air. SOURCE: American Images, Marshfield, Wisconsin.

private storage companies under contract to petroleum and other natural-resource companies, environmental and engineering companies, and, to a much lesser degree, museums, university departments, and various municipal agencies. Most core facilities are owned and managed by the owner of the core; however, some cooperative ventures have proven successful. C&M Storage Inc. in Schulenberg, Texas, is an example of one of the largest such facilities (see Sidebar 3-1). It houses cores and samples from 65 companies and operates as a shared rental facility.

Government's Current Role

There are no state or federal requirements for the collection or retention of core or cuttings from wells drilled on public lands for oil, gas, or mineral exploration or research. DOE, which has several major drilling projects, such as the one at Yucca Mountain, has no formal policy dealing with the deposition or retention of cores. Most of DOE's research and development is performed by contractors whom DOE may ask, on a case-by-case basis, to ensure that cores and other data are maintained and publicly available (Edith Allison, DOE, personal communication, 2002). The U.S. Army Corps of Engineers is required to retain core for a fixed length of time (see chapter 2), after which the risk of loss is high.

The U.S. Nuclear Regulatory Commission (USNRC) uses geoscience data in a variety of ways. This includes evaluating data collected by applicants, licensees, and their contractors, who submit geoscience information to the USNRC to support proposed licensing and decommissioning activities. USNRC staff and contractors also conduct independent sampling, testing, and analyses to confirm information submitted by licensees, to provide guidance, and to develop regulations in accordance with U.S. laws and policies. The USNRC retains geoscience data and information included in licensing documentation submitted for docketing, such as maps, imagery, trench and borehole logs, geophysical and seismological measurements, data sheets from modeling and analyses, field notebooks, and reports. However, the USNRC has no facilities to store and does not retain physical geoscience data from licensees, such as drill core and cuttings, rocks, mineral and water samples, and specimens used in lab tests. Applicants and licensees may be required under USNRC regulations to maintain documented results of tests, analyses, and evaluations and to retain geoscience data and collections. Retention varies from a specified period to the lifetime of the facility (e.g., Code of Federal Regulations Title 10, Part 50, Appendix B, Quality Assurance [QA] Criteria for Nuclear Power Plants and Fuel Reprocessing Plants). The USNRC's Center for Nuclear Waste Regulatory Analyses (CNWRA), which is an independent, federally funded research and development center supporting USNRC's high-level radioactive waste regulatory program, also is required by the USNRC to follow Part 50, Appendix B, QA requirements. Other USNRC contractors and consultants such as national laboratories, geotechnical and groundwater sampling and testing companies, or the USGS may store or preserve geoscience materials at their discretion or under their respective organization's requirements, if any (Philip Justus, USNRC, personal communication, 2002).

The National Science Foundation (NSF) requires cores to be retained by NSF-funded drilling projects, and the NSF Division on Earth Sciences has a general policy on preservation (NSF/EAR, 2002; see Appendix G) as does U.S. Global Change Research Program (USGCRP, 1991). Unfortunately, item number 8 of NSF/EAR's general policy allows decisions on repositing and retaining geoscience data and collections to be made by a single person (the program officer) within the foundation. Requiring principal investigators to report disposition of federally funded geoscience data and collections, and requiring external reviewers and review panels to evaluate this aspect of previous research, would ensure that data and collections would be accessible to the general public. Ice cores collected with funding from NSF's Office of Polar Programs enter into the public domain timed on a project-by-project basis (NICL-SMO, 2000).

The USGS Core Research Center in Lakewood, Colorado, which houses core from 31 states, is the only national repository for publicly accessible core in the United States (see Sidebar 3-2). Unfortunately, the staff must discourage or turn down offers of many collections because of space limitations and inability to absorb the additional workload (Tom Michalski, USGS, personal communication, 2001).

SIDEBAR 3-2
USGS Core Research Center at the Denver Federal Center, Lakewood, Colorado

Founded in 1974, the USGS Core Research Center houses approximately 1.1 million feet of core from 31 states, approximately 95 percent of which was donated by petroleum and mining companies. It currently houses the entire state collections of Colorado, Montana, and Wyoming (with no compensation for doing so), as well as other federal agencies and universities. The facility also houses 15,000 thin sections and 50,000 well cuttings from collections from 27 states. The collection represents 44,507 miles of drilling with an estimated replacement cost of at least $10 billion (NRC, 1999a). The center staff of three serves 1500 to 2000 visitors annually. The center has an annual budget of $275,000 for salaries, benefits, and operating expenses, and pays $550,000 annually in rent. Decisions on accession and deaccession of geologic material are made by the manager of the facility, with input from USGS scientists.

Although the USGS core facility serves a very important purpose, under-funding and limited remaining storage capacity (10 percent) are ongoing concerns. Indeed, in 1995, the available space at that time was reduced by 40 percent. The center also is understaffed. Following USGS's 1995 reduction in force, the staffing level has declined from eight to three full-time employees (Tom Michalski, personal communication, 2001). Nonetheless, the facility serves as a vital resource for industry, federal, and university scientists.

Committee Conclusions of Best Practices: (1) state, federal, and private collections; (2) relatively large and complete regional holdings; (3) good examination and screening space; (4) good clientele support.

The Committee visited the USGS Core Research Center in June 2001.

Two government-funded projects that provide excellent models for preservation of cores in the public sector are the Ocean Drilling Program (Sidebar 3-3) and the National Ice Core Laboratory (NICL) (Sidebar 2-11). The latter is housed in the same large building as the USGS Core Research Center in Lakewood, Colorado (Sidebar 3-2).

Many, but not all, state geological surveys also maintain core repositories of varying size. The cores in these facilities usually are acquired as a result of regulatory compliance on the part of resource companies active in the state, or via donations. Most geological surveys acquire geoscience data and collections from their state alone, although some acquire regional data. Responses to the committee's questionnaire (see Appendix B) indicate that these repository facilities vary from climate-controlled warehouses to cargo containers on gravel pads. The facilities of the Bureau of Economic Geology (BEG), University of Texas, represent the largest of the state geological surveys (Sidebar 3-4). Of particular note is the success of integrating two geographically disparate facilities into one management structure. This distribution resulted from Shell Oil's donation of its Midland facility to the BEG, along with its contents (Sidebar 2-2). Such a model serves as a viable public–private partnership for future transfers of geoscience data and collections to the public sector. Another example of successful partnerships is the Alaska Geologic Materials Center, which serves as the state repository and operates under memoranda of understanding with a range of government agencies to preserve of geoscience data and collections (Sidebar 3-5).

Summary of the State of Cores and Cuttings

Until recently, most large petroleum companies held their cores and cuttings in large company warehouses. However, with increasing pressure to search for extractable resources in offshore and international settings, along with a trend toward mergers and consolidations, many have begun to consider disposing of domestic geoscience data and collections to save costs. For example, a number of transfers of cores from industry-owned storage to other repositories already have occurred (see Table 2-2). In other instances, cores and cuttings simply have been discarded (see for example Sidebars 2-3 and 2-5). The sheer bulk of rock cores in particular is the main threat to their preservation. They occupy space and are difficult to move cheaply.[1] It is in the industrial sector that large numbers of cores and cuttings are at the greatest risk of being lost.

The ODP and the NICL are examples of scientific communities coming together and working with federal agencies to preserve cores from their respective scientific disciplines of oceanography and glaciology. However, long-term sources of funding to maintain these and other repositories are ongoing concerns. The rock-core community offers examples of cooperative efforts as well (e.g., C&M Storage, BEG), but these tend to be isolated instances. Some government repositories that hold rock core are under-funded or under-staffed (e.g., USGS; see Sidebar 3-2), or have no policy for retaining core in the long term (e.g., USACE; see Chapter 2), or have core at risk of being lost (e.g., DOE; see Sidebar 2-10).

MEDIA CONTAINING SUBSURFACE DATA

Data collected from below Earth's surface can be divided into data collected from or associated with drilled wells, and data acquired by other means. Data from wells are commonly recorded as well logs or geophysical logs— paper or electronic records of measured observations or tests made on the rocks through which the drill passed— and include measures of an increasingly large array of physical parameters (see Table 3-2). These tests reveal much about the nature of the rocks that might not be apparent from cores or cuttings. Seismic data result from sending vibrations (produced by explosions or mechanical devices) into Earth. Different layers beneath the surface reflect these vibrations back to the surface in different ways, which allow scientists to develop a picture of Earth's structure below its surface across wide areas.

Although much of the early subsurface data is of lower quality than data gathered with new technologies today, they are very useful for helping to plan a more efficient collecting strategy using new techniques. In other words, old logs and seismic tapes can be used to determine whether additional cost, time, and effort are required in a given area, or whether that area clearly has no potential for resources. (Resources would include such things as minerals, oil, water, clay, sand and gravel, limestone, and even diamonds.)

The majority of subsurface data was and is collected by industry, with smaller but still significant amounts collected by government and academic researchers (see Table 2-1). For example, the latter groups collect a majority of their seismic data as a result of earthquake activity, but use the same collectors and processes to determine the occurrence of explosions triggered by underground bomb tests.

Various collections of raw and test data such as seismic, well log, and petrophysical data, including porosity and permeability tests, are created and held by petroleum companies, environmental and engineering companies, geological surveys, federal agencies, individual researchers at colleges or universities, and private data-storage companies. Some of the largest holders of these data collections include IHS (Information Handling Services), formerly Petroleum Information/Dwights, a commercial operation based upon the reuse of geoscience data (IHS Energy Group, 2002). Their data are used widely by groups in the petroleum industry, govern-

---

[1]The cost of inventorying and physically handling cores during a move varies, but $10 per box is typical (Robert Shafer, C&M Storage Inc., personal communication, 2001).

SIDEBAR 3-3
Ocean Drilling Program Facilities

The Ocean Drilling Program (ODP) is an international scientific drilling endeavor sponsored by the U.S. National Science Foundation (NSF) and 21 participating countries. The prime contractor for the program is Joint Oceanographic Institutions (JOI), Inc., a private, non-profit corporation based in Washington, D.C. JOI Inc. was established in 1976 to manage cooperative research programs for the international oceanographic community under the oversight of a consortium of 14 U.S. academic and research institutions.

Data gathered by the ODP are proprietary to the members of the appropriate drilling leg scientific party for 1 year after sample collection and are then released to the public domain. The ODP and its predecessor, the Deep Sea Drilling Program (DSDP), store cores in four repositories. The Gulf Coast Repository (GCR) at Texas A&M University in College Station maintains more than 140,000 meters (459,318 feet) of ODP core obtained from the Indian and Pacific Oceans. Its operational costs in fiscal year 2001 were $152,204. The West Coast Repository (WCR), at Scripps Institution of Oceanography in La Jolla, California, maintains 130,960 meters (429,659 feet) of DSDP core from the Indian and Pacific Oceans. The WCR was funded for $147,527 in fiscal year 2001. The East Coast Repository (ECR), at the Lamont-Doherty Earth Observatory in Palisades, New York, maintains more than 80,000 meters (262,467 feet) of ODP and DSDP core. The ECR was funded for $261,467 in fiscal year 2001. The Bremen Core Repository (BCR) at the University of Bremen, Germany, maintains more than 72,000 meters (234,000 feet) of ODP cores obtained from the Atlantic and Southern Arctic Oceans. Curation costs for the Office of the Curator, which oversees all the ODP repositories, were funded at $133,030 in fiscal year 2001. Storage and maintenance of ODP material will continue through fiscal year 2004 when these materials will be transferred to the Integrated Ocean Drilling Program (IODP). The facilities have planned to have enough space to accommodate additional cores from the next 2 years of drilling, and will maintain the cores an additional year, to allow the IODP time to arrange their storage plans. Refrigerated storage at the GCR is shown below.

The Joint Oceanographic Institutions for Deep Earth Sampling (JOIDES) science advisory structure is responsible for the provision of scientific advice and guidance to ODP management. The science advisory structure is composed of the JOIDES Executive Committee and a number of scientific and technical advisory committees and panels (see Figure 4-1). The advisory structure office (the JOIDES Office), which rotates between U.S. and overseas institutions at 2-year intervals, currently is located at the Rosenstiel School of Marine and Atmospheric Science, University of Miami. Texas A&M University, as the program's science operator, manages the drillship (JOIDES *Resolution)* operations, shipboard staffing, data collection, core curation, and publications. The Borehole Research Group at Lamont-Doherty Earth Observatory is responsible for providing downhole geophysical logging services, such as collecting, processing, and distributing logging data.

Committee Conclusions of Best Practices: (1) wide (geographic) and diverse clientele; (2) community- and user-based science advisory committee; (3) common-sense regional repositories with good, regionally based holdings; (4) private, state, and federal consortium; (5) research-community emphasis on timely publication of results from collections studies and citations of collections use; (6) adequate funding (as of 2001).

SOURCE: Frank Rack, JOI, personal communication, 2001.

Interior of the Ocean Drilling Program GCR, College Station, Texas. Each box contains a partial length of a single core. SOURCE: Ocean Drilling Program.

SIDEBAR 3-4
Bureau of Economic Geology, University of Texas at Austin

The BEG Core Research Center (see photograph below) is a large state repository containing more than 1.2 million boxes of core and cuttings. Two facilities administered by the University of Texas comprise the center: one in Austin (93,000 square feet) and one in Midland (45,500 square feet). The buildings include research facilities for study and observation. The collection is growing by about 2,000 boxes per year, but three large donations (626,000 boxes) substantially increased the volume very quickly. The core repositories, which cost about $350,000 per year to operate, employ two full-time staff members in Midland and four in Austin. The Geophysical Log Facility in Austin contains more than 800,000 geophysical logs and is growing by 14,000 logs per year. The committed 6,085 square feet of space will have to be reorganized to accommodate growth. The cost of operation for the Geophysical Log Facility is $150,000 per year. State law mandates maintenance of these facilities.

The BEG estimates that 80 people per month request logs and about 400 people annually use the core repository. Most of the inventory is catalogued, but it is difficult to keep up with the influx of data. Data are acquired through donations and by official record submittal as required by law. Data rarely are refused, but available space is declining and currently stands at about 10 percent. Funding comes from a state-appropriated account, but other funds come from American Petroleum Institute, DOE, and endowment funds established by Shell Oil Company to provide care for their large donated collection (see Sidebar 2-2).

BEG is an excellent model of partnerships serving regional needs. It combines state funding with federal grants and donations from private industry to preserve and make accessible geoscience data and collections to the public. The donation from Shell Oil also enables the company and others to maintain access to their geoscience data indefinitely at little or no additional cost.

Committee Conclusions of Best Practices: (1) state, federal, and private support; (2) endowment for some parts of holdings; (3) on-line information about some holdings; (4) good examination and screening space; (5) large, relatively complete holdings of regional importance; (6) state mandate and support to maintain facilities.

The committee visited BEG in August 2001.

SOURCE: George Bush, Sigrid Clift, William Fisher, Daniel Ortuño, Douglas Ratcliff, and Scott Tinker, Bureau of Economic Geology, University of Texas at Austin, personal communication, 2001.

Bureau of Economic Geology, University of Texas at Austin. Flexible-space shelving like this allows storage of cores (right third of photograph), cuttings (left two-thirds of photograph), and other items of various shapes and sizes. SOURCE: David Stephens, BEG, University of Texas at Austin.

> **SIDEBAR 3-5**
> **Alaska Geologic Materials Center**
>
> The Alaska Geologic Materials Center (GMC), Eagle River, Alaska, retains geologic materials from industry and state agencies. In addition, the GMC has agreements with the BLM, USGS, and MMS to archive their rock materials. The GMC holds collections of cores and cuttings from 1,250 oil and gas wells, 920 holes representing 145 mineral prospects and developments, 4 wells representing a geothermal prospect, and 12 holes representing two proposed dam sites on the Susitna River, in addition to a range of other geologic information. A 1982 agreement between the Alaska Department of Natural Resources and the USGS established the GMC. In the original agreement, the value of the collection was stated as "hundreds of millions of dollars." Core and other samples from the National Petroleum Reserve in Alaska were among the initial samples preserved in the GMC. Catalogs of all materials are available on computers at the center, though not on the Internet.
>
> The facility has 9,000 square feet of heated space, but also makes extensive use of large unheated transport (CONNEX) containers for storage of materials, and can expand by purchasing additional containers and shelving. Costs of operating the GMC (a minimum of $110,000 per year) are borne primarily by the state, with additional support from industry donations. Major upgrades have been achieved using federal funds ($300,000 in 1984 from the USGS, and $460,000 in 1999 from the BLM). In 2000, 72 percent of users of the GMC were from industry (oil, gas, and coal), 10 percent were from government agencies, and 18 percent were from academia and the public. No user fees are charged, but clients from industry must cover costs of processing materials.
>
> Committee Conclusions of Best Practices: (1) state, federal, and private holdings; (2) state, federal, and private support; (3) some cost-recovery from industry users; (4) relatively complete holdings of regional importance.
>
> SOURCE: John Reeder and Debbie Patskowski, GMC, personal communication, 2001, 2002.

ment agencies, and others. The Incorporated Research Institutions for Seismology (IRIS, 2002) is an example of a consortium approach for seismic data retention, assimilation, and use.

Government's Current Role

Government regulatory agencies frequently require filing of at least some subsurface data from oil and gas exploration (though not cores or cuttings). The U.S. Bureau of Land Management (BLM) Fluid Minerals Division, for example, requires deposit of ***completion records*** and logs for wells drilled on federal lands. These data reside in approximately 50 BLM offices around the country and are accessible only within the BLM. They are part of the Automated Fluid Mineral Support System, but there is no overall index for this system (Duane Spencer, BLM, personal communication, 2001).

The BLM Solid Minerals Division is responsible for coal, uranium, and other leasable solid mineral exploration on federal lands. This does not apply to lands acquired by "claim" (non-leasable or metallic minerals). According to statute (43 Code of Federal Regulations [CFR] 3484.1(a)(4)), the lessee "shall retain for 1 year, unless a shorter time period is authorized by the authorized office, all drill and geophysical logs and make logs available for inspection or analysis by the authorized officer, if requested." The "authorized officer, at his discretion, may require the operator/lessee to retain representative samples of drill cores for 1 year." There is thus no requirement for permanent data storage of any type. According to the BLM, "a database for coal lease and reserve information called the Solid Leasable Minerals System was developed in the mid-1980s, but discontinued in 1995 due to telecommunications problems and issues concerning the protection of confidential data" (James Edwards, BLM, personal communication, 2001). The USGS uses BLM data in its assessments of national coal resources. To complete the most recent assessment (in 1999), USGS staff commonly traveled to individual BLM offices and obtained hard copies of maps of coal outcrops and mines on BLM lands because limited digital data were available at that time (Suzanne Weedman, USGS, personal communication, 2002).

The Minerals Management Service (MMS) requires the submission of completion records in hardcopy form for oil and gas wells drilled on the continental shelf (pursuant to the Outer Continental Shelf [OCS] Lands Act, as amended [43 U.S. Code 1331], and MMS regulations 30 CFR § 250, 30 CFR § 251, and 30 CFR § 280). Basic well log information is kept for 2 years and 60 days, or until the lease expires, whichever comes first. Beginning in 1976, seismic survey data are held for 25 years before being released. The first

TABLE 3-2 Physical Parameters Recorded in Well Logs

Well Log	Parameter(s) Recorded, Interpreted, or Inferred
Spontaneous potential (SP)	**Formation** water resistivity; to detect permeable beds and **bed** thickness, shale **lithology** and amount
Resistivity, deep	True formation resistivity; to determine water and hydrocarbon saturation of a formation away from the mud invaded zone
Resistivity, shallow	Formation resistivity of the mud-invaded zones; to define bed boundaries
Resistivity, microresistivity	Formation resistivity of the drilling mud-flushed zones; to delineate permeable beds and their thickness
Dipmeter, microresistivity	Direction and angle of formation dips, structural and **stratigraphic** relationships of formation bed; for accurate definition of bed thickness and boundaries
Borehole caliper	Diameter of the borehole; location of porous and permeable zones
Natural gamma ray	Lithology and volume of shale
Induced nuclear radiation: Gamma-gamma Neutron Pulsed neutron (cased hole log) Electromagnetic propagation (EPT)	Bulk density; to deduce formation porosity Hydrogen content; to deduce formation porosity Water saturation; to monitor production performance over time Saturation of flushed zone and percentage of water-filled porosity
Nuclear magnetic resonance (NMR)	Permeability and pore-size distribution, water held in clays and in fine pores; determine moveable fluids and complex lithology
Acoustic (sonic)	Travel time of a sonic wave through a formation; to deduce porosity, fractures, and **vugs**, seismic calibration, **geopressure tops**; rock consolidation, integrity of cement bond between pipe and the formation
Temperature	Formation temperature; to detect producing gas zones and fluid injection intervals
Borehole televiewer	TV picture of a borehole

SOURCE: Adapted from Bradley, 1987; Serra, 1984.

release occurred in 2001 (Gary Lore, MMS, personal communication, 2001).

The National Archives and Records Administration (NARA) does not generally accept raw and test data[2] (such as subsurface data), given NARA's current level of support in this area (NARA staff, personal communication, 2001). Nonetheless, a previous report on preserving scientific data noted specifically that "…a coordinated effort involving NARA, other federal agencies, certain nonfederal entities, and the scientific community is needed to preserve the most valuable data and ensure that they will remain available in usable form indefinitely" (NRC, 1995a, p. 32).

In addition to federal requirements, states have various requirements for submission of data, as well. In Oklahoma and Kansas, for example, submission of only paper well logs is required. In Wyoming, digital submission of well logs is encouraged, but not compulsory (WOGCC, 2001a; see Sidebar 4-5).

---

[2]However, NARA currently holds field notebooks from the Coast and Geodetic Survey containing topographic, hydrographic, astronomical, magnetic, or seismic data, depending on the particular survey. In total, NARA holds 3,367 linear feet of these records (committee survey response, 2001).

In contrast to the United States, some other countries have aggressive policies concerning public deposition and retention of many subsurface data (and cores) acquired from government and public lands (Sidebar 2-5).

Summary of the State of Subsurface Data

While, in terms of absolute numbers, the challenges related to preservation of subsurface data loom large, the situation is not so dire as might be expected. Some of these data are of such immediate- and short-term economic value that an entire industry has formed around its creation, re-sale, and use (for example, IHS and Veritas DGC, Inc. [Veritas, 2002]). The challenge lies in preserving data that, while still having value, pose difficulties for preservation because of their format or lack of current economic interest. Many of the older paper logs and tapes with seismic data fall into this category.

PALEONTOLOGICAL COLLECTIONS

Fossils are the remains or traces of living organisms from the geological past preserved in Earth's crust. They include a huge variety of objects ranging from dinosaur bones and

TABLE 3-3 The 17 Largest Fossil Collections in the United States[a]

Entity	Holdings by Amount (million specimens)
National Museum of Natural History	31
Virginia Museum of Natural History, Martinsville[b]	10
University of California Museum of Paleontology, Berkeley	5
Peabody Museum of Natural History, Yale University, New Haven, Connecticut	4.5
American Museum of Natural History, New York, New York	4
Texas Memorial Museum, University of Texas, Austin	3.8
Los Angeles County Museum of Natural History, Los Angeles, California	3.5
Paleontological Research Institution, Ithaca, New York	3
Florida Museum of Natural History, University of Florida, Gainesville	2.6
Burke Museum of Natural History, University of Washington	2
University of Michigan Museum of Paleontology, Ann Arbor	2
Field Museum of Natural History, Chicago, Illinois	1.3
U.S. Geological Survey Paleontological Collection, Lakewood, Colorado	1.2
Academy of Natural Sciences, Philadelphia, Pennsylvania	1
Museum of Comparative Zoology, Harvard University, Cambridge, Massachusetts	1
University of Iowa Paleontological Collection, Iowa City	1
University of Kansas Paleontological Collection, Lawrence	0.8

[a]Estimated number of specimens at about the year 2000. No major natural history museum actually knows how many fossil specimens it has. Most institutions estimate their holdings by counting or estimating "*lots*" (a "lot" is a set of specimens collected in one place at one time, and can contain 1 or 10,000 specimens), and then using an average number of specimens per lot to estimate total number of specimens. The estimates in this table are based largely on a survey of major collections conducted in 1996 and updated in 1999–2000 (as reported in White and Allmon, 2000). Anecdotal evidence suggests that they may be incorrect by as much as 20 to 30 percent.

[b]This number is based on data on the VMNH website (VMNH, 2001). Most of these collections were transferred to VMNH from the USGS in Reston, Virginia, in 1996. This number is so much higher than those of other institutions that it suggests that a different technique was used in estimating.

footprints to petrified wood to impressions of shells on large rock slabs to the remains of single-cell organisms mounted on microscope slides. Fossil collections conventionally are categorized by the type of organism (vertebrate, invertebrate, plant, microfossil) and organized either by type of organism (systematic or taxonomic collections) or by age (stratigraphic collections). Collections of fossil bones or trackways of animals may consume large amounts of space, whereas collections of microfossils mounted on slides typically occupy much less space.

Commercial trade in fossils has increased considerably in recent years. Large collections are now frequently assembled by individuals via purchase, and a large volume of fossils is held and handled by full- and part-time fossil dealers (see e.g., Morell, 1998).[3] Institutions typically purchase fewer fossils than they acquire by other means, but when they do it is usually a single, special specimen (or a few of them) for special purposes, such as an exhibit, and for which funds have been specially donated. Few museums have acquisition budgets or space for the regular purchase of fossils.

Fossils have been, and continue to be, collected for a variety of reasons, including industry's exploration for fossil fuels, geological mapping, and basic research into the history of Earth and its life. Fossils are collected by exploration geologists looking for mineral resources or making geological maps; by college and university faculty and museum curators pursuing research on topics from the history and evolution of life to global climate change; by undergraduate and graduate students in the pursuit of their studies, especially in geology and biology; and by amateurs for recreation or self-education. In all of these cases, fossil collections serve as the archives and reference sources for such activities.

Fossil collections are held by museums (state and federal government, college and university, private and semi-private), geological surveys (federal and state), colleges and universities, and private individuals (see Table 3-3). Many petroleum companies previously held fossil collections, but most of these have been transferred to museums or universities over the past decade, largely as a result of the general de-emphasis on basic research in industry, combined with increasing use of outsourcing to consultants for industrial paleontological work.

No studies have documented the strong suspicion of many paleontologists that the number of fossil collections being orphaned in the United States is increasing. Before the 1980s, orphaned collections were not widely discussed. Even after

---

[3]The issues surrounding the commercial trade in fossils, and collecting of fossils for sale, particularly vertebrate fossils and especially those from public lands, are complex, contentious, and controversial. Recent accounts include Marston (1997), Davidson (1999), Simpson (2000), McFarling (2001), and Toner (2001).

the issue became widely discussed in the paleontological community in the mid-1990s, it remained difficult to quantify. Recent surveys suggest that more than half of the largest American collections contain adopted orphans acquired within the last 5 years. These orphaned collections represent millions of specimens (White and Allmon, 2000; see also Table 3-4).

Government's Current Role

The federal government's current role in managing paleontological collections falls into four categories: responsibility for collections made by federal agencies on federal lands; regulating fossil collecting by the public on federal lands; support for non-federal collections (via the NSF); and mandate and funding for the USGS and the NMNH (see Sidebar 2-9).

Federal Collections

The issue of caring for collections made on federal lands under federal auspices is broader than just fossils; it includes consideration of materials ranging from Native American artifacts to plants to zoological specimens. How to manage these federal collections has been the subject of considerable recent attention from federal land managers, museum curators, and professional scientists (for a summary, see Faul-Zeitler, 1998). Many issues remain to be resolved. As far as fossils are concerned, although neither the USGS nor the NMNH currently perceives an obligation to house or care for all fossils collected by federal agencies on federal lands (see Sidebar 2-9), there is increasing interest on the part of other agencies—such as the NPS, USFS, and BLM—to care for fossils collected on lands under their respective jurisdictions (see, e.g., Sledge, 1998). There have also been some recent efforts to coordinate consideration and solution of fossil management issues among several of these agencies (e.g., informal interagency working groups and sessions at professional meetings).

Collecting on Federal Lands

Considerable public, scientific, and legislative debate has taken place over the past decade about the regulations covering collecting of fossils by the public on federal lands (Department of the Interior, 1999; Pojeta, 2000; Secretary of the Interior, 2000). This issue frequently is linked to discussions of the commercial trade in fossils, especially those collected from public lands. Some opinions strongly support the notion of somehow regulating access at least to rare paleontological materials found on federal lands. Equally strong opinions counter that if non-professional paleontologists (including commercial collectors) are not allowed to collect freely on federal lands, many valuable fossils may be lost to science. It is too early to forecast a national consensus on this contentious issue.

Support for Non-Federal Collections

For several decades the National Science Foundation has provided modest support for curation and storage of fossil collections at museums, colleges, and universities. Since 1998, NSF has provided approximately $6 million in grants for support of paleontological collections. This has been divided among three programs in biology. The Biological Research Collections program has provided $3,368,456 for direct support of paleontological collections; the Systematic Biology program has provided $2,202,398 in support of research using paleontological collections; and funds from the Biotic Surveys program, totaling $362,979, have supported fieldwork for contribution of specimens to paleontological collections (Larry Page and Cindy Lohman, NSF, personal communication, 2002). For most non-federal repositories, NSF is by far the single largest source of funds for collections support, including shelving systems, staff, and supplies. The Earth Sciences Program at NSF provides no direct support for geoscience collections (H. Richard Lane, NSF, personal communication, 2001).

Summary of the State of Paleontological Collections

Taken altogether, fossil collections in the United States are probably the largest of any nation (Allmon, 1997). Overall, U.S. collections are probably also among the world's best curated. They have, for the most part, not suffered the ravages of neglect, war, and poverty that have afflicted collections in many other countries (Allmon, 2000). Consequently, U.S. fossil collections are visited and studied by scientists from almost every nation in the world. U.S. fossil collections, however, are at a crossroads. They have grown beyond the capacity for existing repositories and institutions to care for them adequately. Priorities have changed among

TABLE 3-4 Paleontological Collections in the United States at Risk of Becoming Endangered or Orphaned in the Next Decade

Type of Collection Holder	Estimated Number of Specimens
Industry	10 million
Colleges and universities	1 million–5 million
Individuals	6 million–60 million

SOURCE: Data obtained from Allmon, 1997.

some of the organizations (e.g., petroleum companies, many colleges and universities, the USGS) that previously cared for them. Yet their importance has never been greater; they continue to serve as fundamental tools for solving societal problems, such as petroleum exploration and studies of global change.

ROCK AND MINERAL COLLECTIONS

Rock and mineral collections include samples collected in the exploration for natural resources, research, geological mapping, teaching, or aesthetics. Sidebar 3-6 illustrates a striking unanticipated use for a rock collection. Mining companies, museums, state and federal agencies, colleges and universities, and individual collectors have all assembled major collections (see Sidebar 3-7, for example), and some of the sources of these collections are now reclaimed, flooded, or otherwise inaccessible (Paul Bartos, Colorado School of Mines Geology Museum, personal communication, 2001). Ore collections, composed of representative samples of different rock types containing metals and other materials, have long been basic teaching tools in colleges and universities. Unfortunately, many universities, including California Institute of Technology, Lehigh University, Massachusetts Institute of Technology, Northwestern University, and Princeton University have closed their collections (A. Sicree, Pennsylvania State University, personal communication, 2001).

SIDEBAR 3-6
The Merrill Collection of Building Stones

In 1880 the Census Office and the National Museum in Washington, D.C., conducted a study of building stones of the United States and collected a set of reference specimens from working quarries. This collection of stones, augmented with building stones from other countries, was then displayed at the Smithsonian Institution. In 1942, a committee was appointed to consider whether any worthwhile use could be made of the collection. It decided that a study of actual weathering on such a great variety of stone would provide valuable information in future construction projects. In 1948, a test wall was constructed at the National Bureau of Standards site in Washington, D.C. (see the image below).

The wall offers a rare opportunity to study the effects of weathering on different types of building stones, with the climatic conditions being the same for all materials. It offers a comparative study of the durability of many common building stones used in monuments and commercial and government buildings.

SOURCE: NIST, 2001; Razand and Stutzman, 2001.

Building stone exposure and test wall, National Bureau of Standards, Washington, D.C. The wall was developed by D. W. Kessler and R. E. Anderson, September 1951. SOURCE: NIST, 2001.

> **SIDEBAR 3-7**
> **Reno Sales–Charles Meyer–Anaconda Memorial Collection**
>
> The Reno Sales–Charles Myer–Anaconda Memorial collection (also known as the Anaconda Rock and Mineral Research Collection) consists of more than 80,000 rock and mineral specimens documenting geologic information throughout the Butte, Montana, mining district. In the 1880s, Butte was the world's biggest copper producer, as well as a significant producer of lead, zinc, gold, silver, and manganese. In addition, specimens in the collection were assembled by Anaconda Copper Mining Company geologists, documenting their travels to other major mining districts throughout the world. The collection is named in recognition of Reno Sales and Charles Meyer, two Anaconda geologists who assembled the collection.
>
> The collection is unique for its size and particularly because the majority of its specimens were collected in the now-inaccessible underground Butte mines. Not until the 1950s did the Anaconda Copper Company move away from labor-intensive underground mining with the opening of the Berkeley Pit, one of the world's largest open-pit mining operations at the time. Mining ceased in the pit in 1982. Today the pit is longer than a mile, nearly a mile wide, and 1,800 feet deep and filled with water.
>
> The Atlantic Richfield Company (ARCO) bought Anaconda's holdings in Butte in the 1970s, and the collection was moved in 1999 to a newly constructed facility near the campus of Montana Tech, which made the collection available to the research community for the first time. The collection is administered by the Montana Bureau of Mines and Geology, a department of the University of Montana.
>
> The remainder of the story typifies geoscience collections that are poorly documented. A large percentage of the specimens (possibly 80 or 90 percent) is not adequately documented, and therefore is of little value for research. For example, a specimen labeled "silver ore from Borneo" is little more than a curiosity or possibly an educational specimen. With inadequate cataloging and the resultant inability of interested parties to discover the full scope of the collection, the bureau has found it increasingly difficult to invest limited state resources in the collection, however rare or valuable, since it is not being used.
>
> SOURCE: Deal et al., 1999; Shovers et al., 1991.

Government's Current Role

No requirements exist at the federal or state level for repositing samples of rocks or minerals gathered on public lands. Collectors prize choice mineral specimens, and the public generally appreciates them. Nonetheless, non-specialists commonly view rocks and ores, which are frequently the most economically valuable, as pedestrian and of little value. Consequently, they are given low priority when allotting scarce storage space.

The Smithsonian Institution has one of the largest collections of rocks and minerals in the world with more than three-quarters of a million specimens. Its acquisition method is fairly typical of geoscience collections, deriving primarily from donations from other government agencies, industry, and private collectors. NSF's Earth Sciences Program provides no support for maintenance or care of rock, mineral, or ore collections. However, the Office of Polar Programs does provide support for the Antarctic meteorite collection, which is housed at the Smithsonian Institution (Timothy McCoy, NMNH, personal communication, 2002).

Summary of the State of Rock and Mineral Collections

Although rock and mineral collections do not represent a large percentage of the geoscience data and collections at risk, they nonetheless represent one of the most neglected categories in terms of preservation. Few government agencies collect these materials, and decreasing numbers of universities maintain teaching collections. Although NSF provides some funding for the curation of paleontological collections, it typically does not do so for rock and mineral collections.

OTHER DATA AND DOCUMENTATION

In addition to physical specimens and data that may be collected as a result of geoscience research and exploration, essential documentation for geoscience projects also includes a wide variety of materials maintained in many different forms and media. Usually unique, these documentary materials may include maps, photographs, field notes, laboratory notebooks, and reports. These materials add essential value in the analysis of geoscience data and collections by providing the nature and context of the research, the data, and the physical samples created or collected as a result of the project.

If the geoscience research was government-sponsored, the federal agency (or agencies) conducting the research is initially responsible for maintaining all these documentary materials for their immediate business and research needs. The National Archives and Records Administration (NARA) is required by law to analyze the long-term historical or other

> **SIDEBAR 3-8**
> **Examples of Government Holdings of Documentation**
>
> 1) USGS Field Records Library, Lakewood, Colorado
>   The Field Records Collection is an archive of materials created or collected by USGS scientists during field studies and other work in the contiguous United States since 1879. Materials in the collection include: field notes, sketches and maps, aerial photographs, analysis reports, stratigraphic logs, and geologic cross-sections (see USGS, nd).
> 2) USGS Photographic Library, Lakewood, Colorado
>   The USGS Photographic Library archives photographs taken by USGS scientists from the 1870s onward. The collection of more than 300,000 photographs includes earth science subjects, such as earthquakes, volcanoes, and geologic hazards, as well as portraits of USGS personnel and 19th century mining operations (see USGS, nd).
> 3) National Mine Map Repository
>   The Office of Surface Mining (OSM) maintains a National Mine Map Repository (NMMR). The NMMR primarily contains maps of abandoned mines. Table 3-5 shows the current holdings by state. Unfortunately, the NMMR's collection of maps continues to be built only through voluntary or informal arrangements with states and the federal government. Many of the maps that have not been reposited with the NMMR are single-copy, paper-only versions that are subject to catastrophic loss from fires, floods, or other events (see, for example, NRC, 2002 p.79). Coordinated digital archives of these maps and records would minimize their storage risks (NRC, 2002).
>
> SOURCE: NRC, 2002; Office of Surface Mining, 1997.

TABLE 3-5 Holdings of the National Mine Map Repository[a]

State	Number of Maps	State	Number of Maps	State	Number of Maps
Alabama	353	Kentucky	4,587	North Dakota	5
Alaska	2	Louisiana	0	Ohio	7,703
Arizona	927	Maine	541	Oklahoma	731
Arkansas	360	Maryland	558	Oregon	333
California	232	Massachusetts	60	Pennsylvania	11,293
Colorado	7,036	Michigan	10,795	Rhode Island	0
Connecticut	475	Minnesota	3,066	South Carolina	54
Delaware	4	Mississippi	84	South Dakota	751
District of Columbia	0	Missouri	8,456	Tennessee	1,155
Florida	0	Montana	727	Texas	1
Georgia	743	Nebraska	0	Utah	647
Hawaii	0	Nevada	940	Vermont	114
Idaho	577	New Hampshire	230	Virginia	8,283
Illinois	2,670	New Jersey	378	Washington	502
Indiana	2,625	New Mexico	121	West Virginia	45,458
Iowa	2	New York	1,184	Wisconsin	504
Kansas	537	North Carolina	1,598	Wyoming	550

[a]Although its holdings are extensive, the NMMR has many gaps in its collection of maps of abandoned mines because of its voluntary and informal agreements with states and the federal government (NRC, 2002). For example, it is unlikely that the state of Texas has only one abandoned underground mine that has been mapped.
SOURCE: Office of Surface Mining, 1997.

> **SIDEBAR 3-9**
> **Denver Earth Resources Library**
>
> The Denver Earth Resources Library (DERL) is a private collection of petroleum industry–related documents, records, books, and maps, organized and stored in 11,000 square feet of commercial space in downtown Denver, Colorado. Records and documents kept at DERL are largely materials donated by major and independent oil companies and individuals. Data include geophysical surveys (seismic), well records, and completion cards. Access to the library data and materials is by membership, with annual dues, as well as with user fees charged to non-members. Student access for academic purposes is generally at no charge. On a typical day about 30 people use the facilities at the DERL.
>
> DERL is a successful, low-tech example of preservation of geoscience data. Data and records generally remain in the format in which they were donated (paper, film, microfiche, or digital). DERL's consistent use implies a commercial niche in the Denver area for storage and access of regional geologic data of high quality and strategic value.
>
> Committee Conclusions of Best Practices: (1) active, steady clientele support; (2) good regional holdings of paper, fiche, and other physical records.
>
> The committee visited DERL in June 2001.
>
> SOURCE: Kay Waller and Laura Mercer, DERL, personal communication, 2001.

value[4] of these materials and determine how long they should be kept beyond the agency's immediate use. If the materials are permanently valuable, NARA will specify a transfer date and an archival repository. Federal documentary materials may not be destroyed, donated to other repositories, or maintained permanently by the originating agency without the approval of NARA (Yvonne Wilson, NARA, personal communication, 2002). Other representative government-housed collections are discussed in Sidebar 3-8.

In addition to field notes, photographs, and maps, other types of data within the other data category include *scout tickets* (written descriptions of individual drill holes, including whether they produced hydrocarbons or not) and completion records (descriptions of the engineering characteristics of a given well). These kinds of data traditionally have been kept in paper or microfiche (see Sidebar 3-9), but increasingly are being collected and retained digitally.

SUMMARY

Geoscience data and collections are archived in a variety of settings around the country, and are collected by many entities within the government, academic, and private sectors, as well as by individuals. They are retained predominantly because they remain useful, or have potential for being useful. These collections can be bulky, particularly cores, which presents a challenge for retaining materials in general and rescuing those that remain valuable but are at risk. There is no federal government-wide coordination of standards for archiving, accession, or deaccession of federal geoscience materials. Yet there are several examples of difficulty in caring for federal collections with current funding levels. Commonly, the success stories the committee encountered were partnerships that had been established between various sectors. Less often, commercial viability led to archiving some geoscience data and collections. There are no set formulas for partners in successful collaborations: successful partnerships occur between the private and public sectors, between state and federal government, and between academia and government. A common element among all these partnerships is a broad user community sharing a common goal—to preserve and make available useful geoscience data and collections.

---

[4] There is some very broad and general guidance on the appraisal of scientific records in Category 15a, "Scientific and Technical Data" of NARA (2002a). For example, category 15a states, "Generally data selected for permanent retention are unique, accurate, comprehensive, and complete, and they are actually or potentially applicable to a wide variety of research problems." Because these published criteria are so broad in scope, NARA usually works with individual agencies on a case-by-case basis to appraise their scientific records to determine their disposition (Larry Baume, NARA, personal communication, 2002). Additional guidelines on appraisal are outlined in NARA (2002b).

4

Managing Geoscience Data and Collections: Challenges and Practices

INTRODUCTION

A number of steps are necessary for successful preservation of geoscience data and collections. This chapter outlines the practices and challenges involved in these steps. Key to the overall success of any preservation effort is an effective management plan, grounded in sound advice from the user community. The idea of user-community involvement was introduced in chapter 2, in the context of the Ice Core Working Group that advises managers at the National Ice Core Laboratory (Sidebar 2-11). Figure 4-1 illustrates how the user community interacts with other areas of management within the Ocean Drilling Program.

STORAGE OF GEOSCIENCE DATA AND COLLECTIONS

Storage, reduced to its most basic level, is the housing of material. Adequate storage is fundamental to the preservation and accessibility of data and collections. Storage is related to, but separate from, curation, which involves safeguarding, cataloging, and locating material; curation is discussed in the following section. A well-stored set of samples may not be curated, but a well-curated sample will be stored adequately.

The repositories surveyed as part of this study (see Appendix B) exhibited no general standards for data maintenance and storage. Consequently, practices vary widely. Cores, for example, are stored in such diverse settings as secure, climate-controlled buildings with well-built storage racks (e.g., ODP), to unimproved metal shipping containers, to boxes stacked on pallets outdoors where they are exposed to the elements. Some cores require specialized storage and maintenance if they are to remain useful for long periods. Ice cores must be stored at –15°C or below, for example, and unconsolidated water-saturated cores such as those held at the ODP repositories (Sidebar 3-3) and the Minnesota Lacustrine Core Repository should be kept moist in a temperature-controlled environment. Repositories that handle these types of cores generally have facilities adequate for the task.

Approaches to storage and maintenance of seismic and well-log data are equally diverse. Some repositories hold only paper records, while many contain both paper and digital records. The digital files might be stored on either magnetic tape or CD-ROMs, the former in climate-controlled settings to slow their deterioration.

Without exception, storage conditions for any data must ensure the integrity of the data themselves as well as their containers, labels, and other metadata, otherwise the data become useless (see Table 2-6). For example, exposure-related deterioration of the identification tags on hard-rock mineral samples, stored in 1990, outdoors under tarpaulins for several years at the Alaska Geological Survey, rendered the cores useless (committee survey response, 2001). All data must be protected from the elements, although conventionally drilled cores and cuttings can be stored under less rigorous conditions than, for example, magnetic media, deep sea cores, paleontologic samples, or ice cores that require temperature and humidity controls. Paper and digital collections minimally require a climate-controlled environment, which by some estimates costs six times more than standard core storage facilities (Robert Shafer, C&M Storage Inc., personal communication, 2001). Over time, however, even rock samples gradually deteriorate from ***oxidation, desiccation,*** or ***disaggregation***. For example, more than half the original zinc core the Tennessee Division of Geology stored has been lost because of inadequate protection from the elements (committee survey response, 2001).

As a result of the perception that rocks and cores can survive years without much attention, they are often stored temporarily under tarpaulins on pallets where they are exposed to adverse conditions. Unfortunately, temporary may become long-term, often resulting in the deterioration of the coverings and boxes, or at least their identifying labels, at which point the utility of the entire collection is lost. To

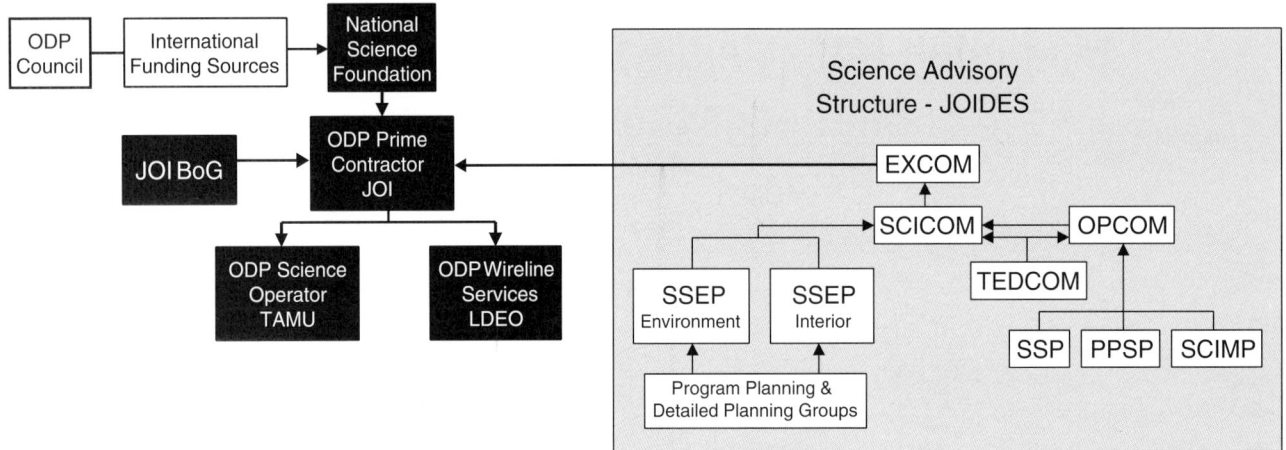

FIGURE 4-1 Ocean Drilling Program (ODP) management structure. Advice from the JOIDES (Joint Oceanographic Institutions for Deep Earth Sampling) science advisory structure (right-hand box) is fed through an Executive Committee (EXCOM) to the prime contractor—JOI. Other acronym definitions for Figure 4-1 are: BoG: Board of Governors; TAMU: Texas A&M University; LDEO: Lamont-Doherty Earth Observatory; SCICOM: Scientific Committee; OPCOM: Operations Committee; TEDCOM: Technology and Engineering Development Committee; SSEP: Science Steering and Evaluation Panel; SSP: Site Survey Panel; PPSP: Pollution Prevention and Safety Panel; SCIMP: Scientific Measurements Panel.

prevent this mistake, the state geological surveys of Alaska (see Sidebar 3-5), Nevada, and Oklahoma have used seagoing shipping containers to store overflow cores until more permanent facilities can be built. Access is limited and not conducive to casual examination, but the vital documentation of sample identity remains intact.

The quality of space provides a degree of security necessary for all collections. While the commercial value of fossil specimens, gems, and meteorites requires that they be protected from theft, all collections deserve protection from loss from other agents of destruction, such as vandalism, weather, insects, mold, and even mishandling by staff and clients. Examples of losses of geoscience data held by state geological surveys extends to earthquake (Alaska), building collapse (Maine), flooding (Kentucky), collapse of shelving (North Carolina and Texas), and exposure (Tennessee) (see Appendix B for sources).

Effective Use of Space

Lack of available space is commonplace at the nation's repositories (see Table 2-3a,b). The quality and amount of space devoted to geoscience collections are highly variable among institutions, and reflect, to some extent, funding and priority assigned by an institution's upper management.

The physical layout of a repository involves several elements relating to space. In addition to space for collections, considerations include adequate processing areas for unpacking, washing, drying, cutting, and sorting samples; cataloging and palletizing; shipping and receiving; workflow considerations from receipt through storage safety and comfort of the staff; security for the collection; and sufficient weight-bearing capability of the shelving and floors (particularly to withstand the load from core collections). For effective use of space, the shelves, racks, cabinets, or drawers in repositories must be closely spaced yet accessible, often stacked high, and durable (so as not to require repeated replacement). Another space consideration is adequate layout and examination space (Figure 4-2), which, ideally, is near the storage area, with appropriate examination equipment (e.g., microscopes), services (e.g., sampling and photography), adequate lighting, and privacy (if necessary).

Ideally, storage facilities are designed to be expanded easily. This is usually a direct function of the value of land upon which the facility is sited. Good examples are C&M Storage in Texas (Sidebar 3-1) and the Ocean Drilling Program repository at Texas A&M University (Sidebar 3-3). The New Mexico Bureau of Geology and Mines repository at the New Mexico Institute of Mining and Technology (NMIMT) constructs additional core storage facilities relatively inexpensively and quickly by erecting 30- by 100-foot, uninsulated, ventilated storage facilities equipped with skylights. Since these are on the NMIMT campus, land-acquisition costs are zero. The recently constructed expandable core curation facilities at the state geological

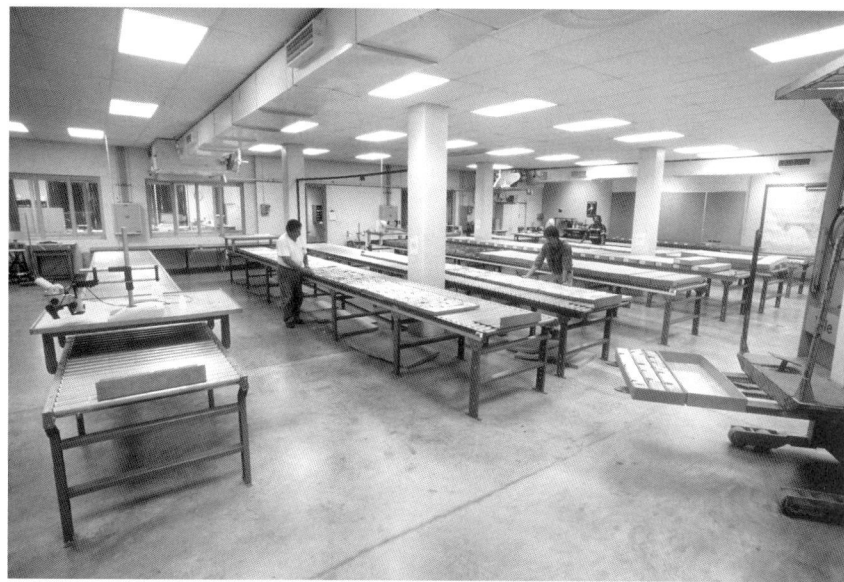

FIGURE 4-2 Onsite study and screening space for core at the Bureau of Economic Geology, University of Texas at Austin. Such space needs to be well-lighted, clean, and climate-controlled. SOURCE: David Stephens, BEG, University of Texas at Austin.

surveys of Kentucky and Ohio also provide multiple use areas for outreach and education.

The facilities mentioned above are in the minority. Most repositories the committee surveyed (Appendix B) are nearly or entirely at capacity (Table 2-3a,b), unable to expand easily, and struggling with old and inappropriate cabinetry for their collections. Innovative actions, however, have allowed some organizations to forestall the need for additional real estate. For example, the National Ice Core Laboratory (see Sidebar 2-11), currently at 90 percent capacity, plans to change its racking system to an adjustable system that will use space more effectively and put the laboratory at 52 percent capacity. Compactor storage, wherein movable racks of drawers ride on rails, saves space and promotes safety and security of the collection. Compactors can postpone more expensive additions to facilities for years. (NSF funds almost all major museum purchases of compactors.) Other space-saving innovations include the use of forklifts with swiveling forks, which allow use of narrower aisles, and therefore a higher density of shelving. Storage space also is saved by trimming and slabbing cores and retaining only a thin slab of the original material. This approach reduced the required storage space by 50 to 80 percent at the USGS facility in Lakewood, Colorado, but many repositories lack the financial resources to fully process all of the core in their collection. Although trimming and slabbing reduces the volume of material to be stored, it is a destructive technique that commonly reduces the types of future analyses that could be performed on the core and thus diminishes their value by some unknown amount. For example, slabbed core is inappropriate for some types of porosity and permeability measurements[1] critical to petroleum engineers and hydrologic modelers, among others.

CURATION OF GEOSCIENCE DATA AND COLLECTIONS

All geoscience data and collections, whether cores and cuttings, rocks and minerals, paleontological specimens, or digital archives, require adequate staff to maintain and curate them in usable condition. Otherwise they quickly deteriorate, become permanently unusable, and ultimately are lost to future researchers. Data curation differs considerably from data storage. Storage in its simplest form is *warehousing*. Curation, on the other hand, is performing the maintenance necessary to safeguard, catalog, and locate samples or records, often bringing them into usable condition through preparation, and keeping them usable for the future. Curation requires protocols for: processing of specimen loans, accession and deaccession (since not everything can or should be archived), promotion of an active research environment to use the collection, ongoing conservation and preservation, and finding long-term funding to ensure their future preservation. Properly curated, the value of a collection will increase through time, and its

---

[1] One reason porosity and permeability measurements are best run on whole core, rather than slabbed core, is the necessity to be able to make measurements on a particular volume of rock, usually a core plug cut out from whole core. Slabbed core does not have adequate volume to permit cutting a plug (Bass, 1992; Morton-Thompson and Woods, 1993).

> **SIDEBAR 4-1**
> **Calgary Core Research Centre**
>
> The Calgary Core Research Centre, in Calgary's University Research Park, is operated as part of the Resources Division of the Alberta Energy and Utilities Board, an energy and utility regulatory agency. The center operates under a legislative mandate to collect, process, and preserve core, drill cuttings, and daily drilling reports from oil and gas wells in Alberta (Oil and Gas Conservation Act/Regulation, Part 11—Well Data, 11.010 to 11.040 and 12.150. This is further specified in Informational Letter IL-OG 76-14). The center also is responsible for providing public access to the material.
>
> The 193,680 square feet of climate-controlled facility serves more than 300 organizations. The staff of 28 manages drill cuttings from 109,202 wells in 236,950 trays (with 56 samples per tray) and core storage for 53,716 wells in 1,047,042 boxes. The center consists of a service and administrative area, research areas, a core repository, a repository for drill cuttings and daily drilling reports, a processing area for drill-cutting samples, and additional patron facilities. It contains 60 core research tables, 7 confidential core research rooms, 50 cubicles for examining drill cuttings, 2 seminar rooms, and 100 equipment lockers.
>
> The Core Research Centre, considered among the best facilities of its type in the world by many who testified to the committee, has been used as a model for many design features of other repositories. For example, features that have been duplicated are the layout space at the Bureau of Economic Geology, University of Texas at Austin (see Sidebar 3-4) (Douglas Ratcliff, BEG, personal communication, 2001), and the forklift system at the Glenside Core Library in South Australia (Elinor Alexander and Brian Logan, Minerals and Energy Resources, South Australia, personal communication, 2001).
>
> The facility is large enough to provide space for another 10 years of core at current accession rates. Of the center's revenues, 70 percent are generated from service fees. The remainder of the budget comes from a combination of the Energy and Utilities Board's well-license fees and the Alberta government (CAD $2.3 million per year [USD $1.4 million], January 25, 2002).
>
> Committee Conclusions of Best Practices: (1) large, well-placed regional facility; (2) very good examination and screening space; (3) cost-recovery allocation; (4) provincial support; (5) large, complete regional holdings; (6) adequate fiscal support (as of 2001).

scientific usefulness will span many decades or even generations (Cranbrook, 1997).

The logistics of handling large quantities of geoscience samples, digital data, and documents are not without staffing and financial consequences. Among the workflow considerations are packaging (creation, standardization, repair), labeling (standardization and formatting), organizing, and moving (loading, unloading, transporting, stacking, or shelving). Repeated handling of specimens must be planned carefully and minimized, if only to conserve staff energy. Each time specimens are handled, the opportunity for spillage, breakage, misplacement, or loss is introduced anew. A model facility for such considerations is the Alberta Core Research Centre in Calgary, Alberta, Canada (see Sidebar 4-1).

Curation involves dedicated and skilled people. Salaries and wages for collections staff are among the largest expense items for most facilities.[2] Consequently, most facilities are short-handed, and curation is concomitantly backlogged. Several facilities utilize innovative means of overcoming staffing shortages by employing part-time student help or volunteers. Typically, volunteers are retired professionals or interested enthusiasts. Reliance on either part-time employees or volunteers can create problems: hiring short-time staff can be difficult because accountability may suffer, work hours can be irregular and unpredictable, often a higher degree of supervision is necessary to avoid errors, and repeated training is necessary to handle turnover. Nonetheless, most museums and other curatorial facilities would be much worse off without dedicated volunteers.

The various types of collections—whether cores, rocks, minerals, gems, fossils, or data—have unique curatorial considerations in addition to the basic curatorial problems of staffing, space, identification, and access. Several case studies, outlined below, illustrate the complexity of curation and the critical roles that staff provide.

---

[2]For example, based on committee survey responses, salaries take up the following percentages of total costs: 30 percent at USGS Core Research Center, 49 percent at Iowa Geological Survey, 25 percent at Denver Earth Resources Library.

Core and Cuttings Collections

Core collections reside in a variety of settings and receive varied degrees of curation and use. The principal curation challenge for core collections is managing their enormous volume and weight. An additional challenge includes archiving core collections in an easily accessible manner. Because lack of space is a constant issue, particularly in public institutions, significant staff time is spent reducing the volume of core collections.

The USGS Core Research Center in Lakewood, Colorado (see Sidebar 3-2), currently has a staff of three and deals with 1,500 to 2,000 users annually. These users are predominantly from the petroleum industry and academia. The center also handles about 1,000 inquiries from people wanting information about the collection each year. With this level of staffing, the USGS can maintain the collection and provide some support services to users, but staff can do only limited processing (slabbing or photography) of new cores. Users needing more intensive processing services must be referred to outside services.

Collections staff at state geological surveys usually range from one to two full-time employees, with additional part-time help (committee survey responses, 2001). Despite budget cuts (e.g., in Iowa and Kentucky), sample collections continue to grow annually at an average rate of about 2 percent. Growth could be greater, but is usually hampered by staff costs or space limitations. Nevertheless, collections staff continue to encourage collections use, while attempting to eliminate curatorial backlogs and encouraging better initial documentation (i.e., better metadata). Geoscience data and collections are used daily at virtually all state geological surveys. The few geological surveys that require collection users to make an appointment do so to schedule access to limited core examination space (e.g., Indiana) or to move boxes because of overcrowded aisles (e.g., Iowa and Kentucky).

Media Containing Subsurface Data

While some subsurface data are in paper format, the majority are electronic, gathered over the years using various techniques and equipment. These data present challenges unique to the electronic environment, such as **data migration** and equipment compatibility. A large volume of seismic data remains on older media such as film or various forms of magnetic tape (see Table 2-1). For example, until the 1980s, seismic data were stored on magnetic tape; now they are routinely preserved on **server farms** (NRC, 1995a,b). The very large volume of **well log** data (Table 2-1) solely in paper form presents challenges related to access and utility, as much as preservation of the medium itself.

Even if subsurface data have been transferred to or already exist in a digital medium, the data are not guaranteed immortality. Data can be lost because of obsolete formats, obsolete equipment, or physical degradation of the magnetic medium (particularly magnetic tape, which degrades more rapidly when storage facilities are not climate controlled or otherwise weatherproof, and should be rewritten about every 5 years). IHS is a for-profit company that makes large investments in migrating old seismic and other data into standardized digital form for distribution to customers (Ron Samuels, IHS, personal communication, 2001). In the long term, data migration and assimilation can add value as the dataset grows. For example, restoration of SeaSat (Appendix F) data demonstrated that constantly reworking data is more cost-effective than ignoring them (NRC, 1998). Digital storage of subsurface information is appropriate because digital data are increasingly being stored in smaller physical spaces with greater cost effectiveness, they can be duplicated and stored in different places, thereby safeguarding them from loss, and they can be accessed and shared more easily in digital format than they can be in a paper or tape format.

As a result of its cost and time-consuming nature, migration of data is a challenge for smaller institutions and organizations that lack the necessary short-term funds. At THUMS (see Appendix F) in Long Beach, California, the seismic data collected for the Wilmington oil field in the 1970s were saved on 1,600-bpi tapes. These tapes were not readable in 1995 when THUMS staff tried to integrate the data with a three-dimensional seismic survey completed that year. Similarly, in 1991, Chevron estimated that 11 percent of its tape data were unreadable because of degraded and outdated storage media (Philippe Theys, Schlumberger Ltd., personal communication, 2001). Low-budget, not-for-profit entities such as DERL (see Sidebar 3-9) have no data migration plan in place, and therefore continue to work with paper, fiche, tapes, and other physical subsurface data storage media. To do otherwise would preclude some users who typically are unable to pay for-profit prices, especially users from smaller companies and academic settings.

Paleontological Collections

All paleontological specimens do not need the same level of curation to be scientifically usable. A hierarchy of curation, described by Hughes et al. (2000) as a curatorial continuum, minimally requires that collections be safe from damage, mishandling, or loss. With an increased investment of staff resources, the scientific value of specimens increases as they progress through the continuum. Typically they are cleaned, sorted, boxed, identified, labeled, cataloged, and perhaps reconstructed. Preparation of fossil specimens involves cleaning and, in some cases, reconstruction of missing portions, which can be extremely time consuming. In the Smithsonian Paleobiology Collection, a cataloger can properly process 15 to 20 specimens per day (committee survey response, 2001).

Typically only a fraction of an institution's collection is brought to a fully prepared state for a specific display, re-

> **SIDEBAR 4-2**
> **National Geophysical Data Center,**
> **Marine Geology and Geophysics Division**
>
> The National Geophysical Data Center (NGDC) in Boulder, Colorado, largely handles digital data collection, storage, and processing. As an indication of the scope and scale of the data storage issues, only the marine geology and oceanographic aspects of the NGDC mandate are described herein; however, environmental data in general are within their charge (see NGDC, 2001).
>
> Marine geoscientific data are stored digitally at the NGDC's Marine Geology and Geophysics Division (MGG). The MGG databases deliver 10 gigabytes of data each month over the Internet. The MGG databases include more than 5 gigabytes of scanned images of marine sediments and rocks, with an additional 2 to 2.5 gigabytes of digital data files, more than 76 gigabytes of multi-beam bathymetry, 2.7 gigabytes of hydrographic data, and 6.9 gigabytes of underwater *geophysical trackline* data. In addition, archived data include more than 3,100 microfilm reels and 30,000 square feet of seismic sections, among many other types of data. These data are accessed by scientists from various government agencies, by academic researchers, and by private citizens. Uses of these data include engineering studies in preparation for laying undersea cables, fish habitat and sea mammal studies, mineral exploration, international mapping studies, and commercial and sport fishing. The NGDC-MGG web site is: *http://ngdc.noaa/mgg/mggd.html*.
>
> Committee Conclusions of Best Practices: (1) excellent on-line accessibility and availability of data and metadata; (2) broad, international user-community involvement; (3) coordinated information flow to and from user community. The committee visited NGDC-MGG in June 2001.

search, or educational purpose. Considering the size of most collections and the expense that would be incurred in fully preparing every specimen, the vast majority of fossil collections are retained unprepared (Hughes et al., 2000).

Budgetary factors also influence the state of sample curation. For example, at the NMNH, budget limitations have prevented critical conservation of the fossil vertebrate and paleobotanical collections (or the replacement of storage cabinets in which they reside) (committee survey response, 2001). In some recent years, the Department of Paleobiology has had no funds at all available to purchase even the most basic supplies, such as specimen boxes.

Rock and Mineral Collections

The curation of rock (including meteorite), mineral, and gem collections poses some unusual challenges. Meteorites are immensely popular among private collectors, so security of the collection is critical to maintaining its integrity and intellectual value. The Smithsonian's National Meteorite Collection currently consists of 22,000 specimens, and it is growing by 50 to 100 specimens annually (committee survey response, 2001). A staff of one or two curators and one or two collections managers curates the collection and facilitates 400 to 500 loans annually for exhibition or research. The Smithsonian's Mineral Collection consists of 500,000 rocks and minerals (including gems). It, too, is managed by a staff of two curators and two collections managers.

Security is a serious concern for both collections. Catalogs of the holdings are not readily available, and electronic access to inventories is viewed cautiously because of fear that publicizing the nature and size of the holdings will compound security problems.

Other Data and Documentation

The curation of paper and digital collections is very much like that in any library. A staff is necessary to accept, catalog, shelve, and maintain the collection to function as intended. For example, the Kansas Geological Survey's Data Resources Library is maintained by eight full-time employees and four part-time (student) employees (Kansas Geological Survey, 2001). In the private sector, the Denver Earth Resources Library uses two full-time employees and two part-time employees to maintain predominantly paper records (see Sidebar 3-9). The library is visited by 30 to 35 self-serve patrons per day.

While holdings of digital data are outside the committee's immediate interest, the following discussion is included to illustrate that staffing needs for handling digital data are not insignificant and should be considered in any plan for improving access to metadata about physical collections. Data holdings at the National Geophysical Data Center's Marine Geology and Geophysics Division require 4.5 full-time employees to manage and maintain the marine geophysical databases (see Sidebar 4-2).

Where digital records exist for incoming data, as is often the case at the NGDC, they are reproduced and held as

TABLE 4-1 Libraries and Geologic Repositories—A Comparison of Cataloging Practices

	Libraries	Repositories
What do they store?	• Printed materials (text and digital) plus audio-visual materials	• Systematics collections (rocks, fossils, etc.) • Cores (rock, sediment, and ice) • Geophysical data (digital, paper, film) • Records (maps, photos, log books, etc.)
Catalog characteristics	• Conform to established international standards • Mostly digital • Interoperable	• Few standards • Some digital/some paper/some microfiche • Little interoperability

SOURCE: Committee survey responses, site visits.

a backup copy. Quality control is performed as data are entered into databases. NGDC has a center-wide metadata entry system following Federal Geographic Data Committee standards (***FGDC***, 2002). As technology changes, data are migrated in new forms and media as necessary. Archived data are in ASCII format, which can be converted to other formats. Contributors inspect and approve any data modified by NGDC before final posting for public distribution.

Data in digital media periodically require refreshing. At the NGDC, the staff continually needs to refresh software, hardware, and training—to protect against media deterioration and technology evolution, and to guarantee accessibility and retrievability.

CD-ROM storage currently is one of the more popular forms of digital data storage. Within the USGS, paper documents, well logs, and seismic displays are scanned into image files and captured to CD-ROM, as are data stored on ***magnetic tapes*** (Linda Gundersen, USGS, personal communication, 2001). Benefits include a simple and low-cost replication process, ability to store multiple datasets (e.g., text, images, video, and audio), and random access of the information.

CATALOGING AND INDEXING

Specimens, samples, or other geoscience data that have no documentation about their origin (metadata) are of little or no scientific value. Materials without such documentation usually are not accessioned into collections and are prime candidates for deaccession efforts when staff time is available. ***Cataloging*** is the process of recording metadata in some centralized database, usually with some kind of index numbering system on index cards, ledger books, or computer software. Table 4-1 summarizes general differences between the state of cataloging in libraries and geoscience repositories.

Cataloging facilitates good management of data and collections, and greatly reduces the cost of using them. Without catalogs many collections are useless, except to the rare expert who knows a specific collection intimately. Cataloging is also necessary to gain a better estimation of the staffing and financial needs for properly curating a collection.

Uncataloged materials are almost impossible to use or loan, and most collections facilities have a backlog of uncataloged materials.[3] At the Smithsonian Institution, priority for cataloging depends on the commercial value of the specimens, the number of specimens acquired per year, and the size of backlog remaining from years in which large USGS or NASA transfers were accepted (committee survey response, 2001). This is especially true where gems, meteorites, or unusual and rare fossils are involved. In the Department of Paleobiology, cataloging primarily is focused on newly acquired ***type specimens*** because of their importance to the scientific community. Within the Smithsonian, the National Museum of Natural History is home to one of the premier geoscience collections in the world. It has an active cataloging program and a database of 5 million records describing 124 million lots of items (collections of fossils or other objects) (committee survey response, 2001). This large number represents just 10 percent of the records required to describe this collection adequately. With the shedding of staff from the Smithsonian's Collection Management Program over the last 10 years, the rate of cataloging has declined significantly. Processing loan requests has been given priority, so that the larger scientific community neither notices nor is affected by the staffing shortage (committee survey response, 2001). The situation at many other museums is much worse.[4] Data on collections of special value (such as type specimens) almost always are available, but data on the great majority of collections are available only by physically examining the paper labels associated with the specimens. In the committee's view, cataloging is an enormous and pressing need for effective use of the nation's geoscience data and collections.

---

[3]For example, new purchases and exchanges at the Los Angeles County Museum of Natural History are cataloged immediately, but old material is catalogued only periodically (committee survey response, 2001). Other examples include, cataloging of cuttings at BEG's Midland facility, which is backlogged (committee survey response, 2001), and the Anaconda Mineral collection (see Sidebar 3-7), of which less than 20 percent is cataloged.

[4]For example, 50,000 of the 3 million specimens held by the Paleontological Research Institution in Ithaca, New York, are cataloged.

> **SIDEBAR 4-3**
> **Profiling the Collections at the Smithsonian:**
> **A Tiered Approach to Collections Description**
>
> The Smithsonian was significantly affected by budgetary cuts throughout most of the late 1980s and 1990s. This resulted in the loss of many critical collections and curatorial staff positions and sharply curtailed other resources necessary for management of the national collections. One result is a backlog of 40,000 volcanic specimens awaiting accession into the Smithsonian's Rock Collection. The USGS gave these specimens in 1995, but until they are curated, they remain available only to selected researchers, rather than to the research community as a whole.
>
> To obtain a better estimation of the staffing and financial needs for properly curating these and its other estimated 124 million lots, the Smithsonian Institution is undertaking a museum-wide collections profiling assessment. Six irreducible factors are being measured: conservation (i.e., physical condition), processing (how much curation is necessary to bring specimens into full museum ownership), storage (from microscopic to building-sized requirements), arrangement (physical and intellectual sorting to provide access), identification, and current status of inventory. Pilot assessments have been performed, and the process is being fine-tuned. The full assessment will provide a means to plan and budget for staff and space needs.
>
> Committee Conclusions of Best Practices: (1) research on and curation of holdings by staff; (2) large, diverse holdings of great national importance.
>
> SOURCE: Sally Shelton, Smithsonian Institution, personal communication, 2001.

The Smithsonian Institution is experimenting with a tiered approach to collections description so that more general descriptions of collections will be available before detailed cataloging is completed. Such an approach might be suitable for other collections, as well. Sidebar 4-3 illustrates an approach currently in progress at the Smithsonian Institution.

Another challenge for users of geoscience data and collections is the lack of any national catalog. Researchers must search out each site or catalog individually and examine it for data or collections of interest. Such catalogs do exist for bibliographic materials in the geosciences, however.[5] Maintained by libraries and archives, these catalogs provide a useful and successful model to follow. A standard system of describing, *indexing*, and formatting the catalog is essential to assist users in locating materials of interest and to allow interoperability among multiple databases and catalogs of materials.

Two of the key characteristics of these catalogs are their adherence to national standards for catalog records (metadata), description of items, and database interoperability (e.g., Library of Congress, 1995), as well as use of widely accepted thesauri and terms for description. Adherence to these standards ensures that users of these catalogs can determine the appropriateness of material for their research or educational needs. These and similar catalogs have proven track records, and garner worldwide acceptance. The invertebrate paleontology community took an important step toward common computerized standards with the development of a common data model (Morris, 2000) that can be used to relate different collections databases to each other.

The committee concludes that *inadequate cataloging is the single biggest inhibitor to productive use of even well-maintained geoscience collections in the United States.* Sidebar 4-4 describes the Institute for Museum and Library Services (IMLS), a government agency that, since 1996, has provided funding on a competitive basis for improving access to information at museums and libraries. Although currently not supported by IMLS, cataloging efforts in the geosciences clearly fall within the institute's mission (Robert Martin, IMLS, personal communication, 2001).

ACCESS

Unless geoscience data and collections are accessible, they are useless. Access to the data and collections themselves, however, is the second step in achieving full access. Access to information about the data and collections (e.g., metadata and catalogs) is the first step in any full-access process.

Before the electronic age, lists of data in collections were kept in serial logbooks or on alphabetical file cards. Someone familiar with the recordkeeping system had to search the data listing and determine the physical whereabouts of the desired data. Access depended on a high degree of institutional memory—that is, individuals who knew the history of the system and who knew and cared about its organization.

---

[5]One example is GeoRef (http://www.georef.org).

> **SIDEBAR 4-4**
> **Institute of Museum and Library Services**
>
> The Institute of Museum and Library Services (IMLS) is a government agency that allocates funds on a competitive basis to museums and libraries for improving access to information. The institute was formed as a result of the Museum and Library Services Act of 1996 (see IMLS, 2002), which moved responsibility for federal library programs from the Department of Education to the institute. The museum program received $28.7 million in fiscal year 1997 and $23.4 million in fiscal year 1999. The library program received $150 million in fiscal year 1997 and $166.2 million in fiscal year 1999.
>
> In 1999 IMLS dispensed $170 million in grants. Grants typically run as long as 2 years. The institute's programs foster the development of digital resources and linkages among and between libraries and museums, and assist museums and libraries in evaluating their programs. The IMLS Conservation Project Support program offers matching grants to museums that identify conservation needs and priorities and perform activities to ensure the safekeeping of their collections. Collections may be in one of four categories: 1) non-living; 2) *systematics* and natural history; 3) living plants; 4) living animals. Grants are available for five broad types of conservation activities: 1) surveys; 2) training; 3) research; 4) treatment; and 5) environmental improvements. In addition, the IMLS offers a Museum Assessment program that supports the assessment of museum operations, collections care, or public service that can result in more effective goals and plans for the museum's future. IMLS has informal partnerships with NSF on initiatives such as the National Digital Library (DLF, 2002) and e-gov (USGSA, 2002).

Archives reliant on institutional memory are prone to degrade when staff members transfer, retire, or otherwise leave the institution. Today, computerized databases of collections holdings can be searched and queried by any number of descriptive parameters, even remotely over the Internet, utilizing much of the same technology developed by libraries. Yet, to a large extent, these systems are not in place for geoscience data and collections.

Traditionally, indices and catalogs have been the means by which researchers learned about new research, data, and collections. Printed indexes of research findings and catalogs of collections were widely available and used for decades. With the advent of the digital age, many of these printed research tools were converted to electronic form (i.e., computerized; see below), allowing easy access and saving time for researchers. Bibliographic databases such as *GeoRef* (AGI, 2002a) and *Oceanic Abstracts* (CSA, 2002), as well as library catalogs, are used worldwide to facilitate the *discovery* of geoscience research.

Unfortunately, the tools for locating geoscience data and collections have not made the same successful transition. In some cases, attempts to keep accurate catalogs of holdings have ceased, while in others, a catalog may exist only onsite.[6] Good tools for locating geoscience data and collections are not absent due to lack of interest. Rather, in many instances, funds to build electronic catalogs and provide Internet access are available only when garnered from various savings in operational funds; new money for these efforts is rarely afforded.

The limited extent to which paper catalogs and metadata have been converted into digital format (or computerized) is a missed opportunity to enhance the use of geoscience data and collections. This is particularly true as digital catalogs, coupled with current Internet technology, have increased tremendously the usage, value, and societal benefits of such holdings (see Sidebar 4-2). Almost all U.S. collections, particularly fossil and mineral collections, are cataloged incompletely, only a few catalogs are available over the Internet, and no comprehensive tools are available to search multiple repositories at one time.

In the United States, *invertebrate* paleontological collections are among the most numerous, but least computerized, of systematic natural history collections. A 1993 survey (Cooley et al., 1993) estimated that they were approximately 8 percent computerized. The USGS paleontological collections are mostly cataloged on paper and as multiple discrete collections (see Sidebar 2-10). While digital catalogs exist for different kinds of materials at the USGS, there is no unified catalog of the USGS holdings (Linda Gundersen, USGS, personal communication, 2002).

At the Smithsonian, as at all other museums, almost no specimens are accompanied by digital data when they arrive (committee survey response, 2001). Specimen data nearly always arrive as donor-generated labels, scientific publications, and maps that accompany the samples. Specimen data

---

[6]In the committee's experience, very few institutions have ongoing programs that add large numbers of catalog entries for specimens/lots that already exist in the collections.

are prepared for computer entry by organizing them on handwritten forms. Although this seems cumbersome, the two-step process minimizes error and leaves a tangible trail. The goal is an error-free collections database. Older digital catalogs exist, but their life expectancy is limited. In the Smithsonian, many digital data reside on main-frame computers. Although such data are not in immediate danger of loss or damage, no transcription program is underway, consequently they remain completely inaccessible even to institution staff (committee survey response, 2001). Data for the Smithsonian's Department of Paleobiology is still managed by use of SELGEM, a database program developed in the late 1960s. Starting in Fall 2002, Paleobiology will use KE EMu,[7] a catalog currently under construction that will incorporate digitized images, documents, spreadsheets, and databases. All SELGEM-based specimen data will be migrated to KE EMu, and data in analog formats that never have been in electronic form will be added manually to records. Once data are in KE EMu, they can be reported in a variety of formats. A significant collateral benefit of computerized catalogues is that they also function as an electronic backup of irreplaceable documents related to the collections.[8]

The situation is somewhat better in other branches of geoscience. The Minerals Management Service is in the process of scanning or digitizing all of its data, and soon it will accept only digital data (Gary Lore, MMS, personal communication, 2001). Other isolated repositories, collections, and projects have been successful in providing digital access to their resources. Some of the best examples include the *Catalogue of Meteorites* (Natural History Museum, London, 2002), Kansas Geological Survey (2001), the National Geophysical Data Center (NGDC, 2001), the National Ice Core Laboratory (NICL, 2001), the Ocean Drilling Program (ODP, 2002), and the Wyoming Oil and Gas Conservation Commission (see Sidebar 4-5).

The Smithsonian's Rock Collection is a unique collection of Earth's crust and mantle rocks. All of its 446,000 lots are computerized, making it the largest curated, completely computerized rock collection in the world. Furthermore, the catalog is accessible over the Internet. Access to most other Smithsonian collections is through curators and collections managers, usually through personal contact via telephone, e-mail, on-site visits, or at professional conferences. With the launch of the Smithsonian's on-line KE EMu catalog, a broader audience will have access to a subset of data about most of the specimens in these collections.

The Internet revolution of data access has major implications for geoscience data and collections. The Internet allows multiple users simultaneous access to data that previously were inaccessible—it does so 24 hours a day, 7 days a week without having to travel to individual repositories. For many institutions, the Internet also is responsible for increased interest in their collections. For example, the average number of hits per month on web pages related to petroleum information at the Kansas Geological Survey increased by 41 percent between 2000 and 2001 (Timothy Carr, Kansas Geological Survey, personal communication, 2002).

Several attempts have been made to improve the level of access to catalogs of geoscience data and collections across the United States. These include the GeoTrek metadata catalog (AGI, 2002b) developed by AGI with funding from DOE. As a prototype for a framework that allows access to digital geologic information, its value lies with the underlying data and the links that permit access to the data. Similar beginnings have been made, albeit on a smaller scale, by the Mines Ministries of several Canadian provinces and Australian states (see Appendix H for examples). Many smaller collections in the United States have benefited greatly from NSF funding for cataloging and ***computerization***. For private entities in the business of providing geoscience infor-

---

> **SIDEBAR 4-5**
> **Wyoming Oil and Gas Conservation Commission**
>
> The Wyoming Oil and Gas Conservation Commission (WOGCC) operates a geologic data management system with minimal staff and a bare-bones budget. Despite these limitations, the petroleum exploration database exists on a real-time, interactive web site (WOGCC, 2001b) that can be accessed from a field location. As a company reports well data from the field, it is saved immediately to the appropriate database and is available on the web. Accessible on the site are 35 databases containing about 12 million records. Using this system, WOGCC staff can issue permits for more than 1,000 wells per month, compared with past rates of about 1,200 per year. The cost to acquire the hardware and software and complete the entry of historic data was $60,000.
>
> Committee Conclusions of Best Practices: (1) digital submission of data and metadata; (2) public access via Internet to real-time data and metadata updates.
>
> SOURCE: Richard Marvel, WOGCC, personal communication, 2001.

---

[7]KE EMu is an electronic museum management system that supports data capture, querying, and museum management functions through a client–server interface. It also includes a web interface for Internet or Intranet access to museum data resources. KE EMu is produced and supported by KE Software, an Australian company. (See http://www.kesoftware.com/Press/release5.html.)

[8]Catalogs of the Department of Mineral Sciences within NMNH should be available on the KE EMu system in May 2002 (Anna Weitzman, NMNH, personal communication, 2002).

TABLE 4-2 Incentives for Improving the Ability to Find Information about Geoscience Data and Collections

- Significantly reduce the time spent searching for information and increase time spent in analysis and use
- Increased timeliness of data availability for emergencies (see for example the opening paragraph of the Executive Summary on Hutchinson, Kansas)
- Increased investment in exploration and extraction of state and national resources (with the attendant advantage of increasing state and federal revenues from taxes thereon)
- Increased use of collections for educational and scientific purposes
- Aid collections management when trying to determine the uniqueness or significance of samples.

mation, an accessible catalog is crucial. Examples include the electronic catalogs of IHS Energy (IHS, 2002) and Veritas DGC, Inc. (Veritas, 2002).

DISCOVERY AND OUTREACH

Discovery entails identifying the existence and whereabouts of desired data and collections in addition to determining their availability, quality, and format. Discovery can occur in a variety of ways: the Internet provides one means of access to computerized records, whereas attribution at the end of journal articles, for example, leads the reader to a source of information. Outreach is another, more assertive form of enhancing discovery that has been applied successfully in the ocean geoscience community (see Sidebar 4-2).

Much can be gained by improving our ability to discover data and collections (See Table 4-2). In a 1992 paper, Blaine Taylor (1992, p. 193) states: "…we simply spend too much time and money searching for, collecting, and pre-processing data before we can even begin the analysis phase of our work. Recent studies[9] indicate that as much as 80 percent of our engineers' and geoscientists' time can be spent in these efforts." The goal of most organizations, whether public or private, is to shorten the discovery time so that the investigator or employee can spend more valuable time actually analyzing the data. Internet-based catalogs allow prospective users of physical samples to determine from afar whether they need to visit a facility, thus saving time. For example, users in Colorado, Texas, and Oklahoma have been able to explore online the holdings of the Wyoming Oil and Gas Compact Commission in Casper, Wyoming (Richard Marvel, WOGCC, personal communication, 2002) (see Sidebar 4-5).

In many instances, however, researchers still gain knowledge of and access to data by traditional means: personal acquaintance, letters, onsite visits, telephone, fax, or e-mail (Wayne Ahr, Texas A&M University, personal communication, 2001). Consequently, data and collections are under-appreciated and under-utilized. More importantly, valuable scientific information that is lost to the discovery process cannot be used for subsequent analyses and interpretations, weakening both. For example, staff in the U.S. Army Corps of Engineers, which has 40 district offices nationwide, have been largely unsuccessful in obtaining funds to publish catalogs of their holdings, leaving their data and collections accessible to the public only with great difficulty (committee survey response, 2001). At the USGS, determining the location, existence, and availability of certain samples may require several phone calls or e-mails (Kevin McKinney, USGS, personal communication, 2001). Access to electronic catalogs of geoscience data and collections is therefore essential to facilitate discovery of these resources by the broadest range of potential users.

Attribution

A basic tenet of science and engineering research is the precept that new discoveries build upon old ones. Scientists are taught to evaluate and acknowledge the research that has come before. This acknowledgement is accomplished through a system of attribution, by citing previous and related work. How to cite previous works is the subject of numerous style manuals and guides to research in all fields of study.[10] It is notable, however, that the style manuals of the sciences rarely refer to or recommend the citation of data or collections (the text in Sidebar 4-6 is one example). This is completely opposite to other areas of study (such as history and the arts) and reflective of the historic reliance of the geoscientists on personal contacts and personal knowledge of collections. The lack of attribution to geoscience data and collections only serves to promote their invisibility and to downgrade their value. The committee concludes that *it is essential for the geoscience community to follow the lead of other sciences and begin to cite (i.e., acknowledge) use of and reliance upon data and collections*. The NRC's suggestion in its overview of NASA's Distributed Active Archive Centers (DAAC) is one approach to this problem (NRC, 1998, p. 40): DAACs are encouraged to post on the Internet a list of publications that make use of their holdings, in a format that would permit an easy search for references with standard web tools. In another example, use of data holdings of the World Data Center for Paleoclimatology is referenced in a standardized manner (see NOAA, 2000).

---

[9]For example, Conoco Inc. conducted a study of all of its geoscientists and engineers in their Exploration and Production organization. The Conoco Task Force asked how much time was being spent on technical work versus data hunting and data management.

[10]For example, *American Chemical Society Style Guide* (American Chemical Society, 1999); *Chicago Manual of Style* (University of Chicago Press, 2002); *Suggestions to Authors of Reports of the USGS* (USGS, 2002).

> **SIDEBAR 4-6**
> **Method of Attribution for Reports Using Ocean Drilling Program Data and Collections**
>
> "This research used samples and/or data provided by the Ocean Drilling Program (ODP). The ODP is sponsored by the U.S. National Science Foundation (NSF) and participating countries under management of Joint Oceanographic Institutions (JOI), Inc. Funding for this research was provided by _____."
>
> In addition, the words "Ocean Drilling Program," "scientific ocean drilling," or "ocean drilling" should be used as one of the keywords provided to journal or book publishers of your manuscripts. This will allow the legacy of the ODP to be tracked by bibliographic databases (e.g., GeoRef).
>
> SOURCE: Frank Rack, JOI, personal communication, 2001.

> **SIDEBAR 4-7**
> **Geoinformatics**
>
> The goal of geoinformatics is to construct multidisciplinary databases to facilitate extraction of knowledge from the geologic record. The geoinformatics community is planning a network through a collaborative research initiative undertaken by a consortium of universities and non-academic partners such as USGS, NOAA, NASA, BP Amoco, BEG, and the Geological Survey of Canada. Earth and computer scientists aim to establish a seamless and integrated network system of geoscience data with software tools for access, analysis, visualization, and modeling. The goal of the Geoinformatics Initiative is to develop a national infrastructure of databases and tools for earth science research.
>
> SOURCE: Geoinformatics Network, 2001.

Outreach

Some of the most exciting discoveries occur through interdisciplinary research, which by its very nature, requires researchers to work beyond their normal boundaries. Consequently, data and collections managers should reach beyond their traditional user communities to educate new users about the existence and utility of the geoscience data and collections they hold. This implies that the organizations archiving these data will have to engage in a certain degree of marketing. For example, NGDC (Sidebar 4-2) promotes its holdings via e-mail and the Internet, through mass mailings, at professional meetings, and with posters and CD-ROMs.

The Internet serves as an effective outreach tool, for example, by making available a wide selection of images of gem, mineral, rock, ore, meteorite, and fossil specimens, as well as related documentation (e.g., field notebooks, historic illustrations). Many museums, including the Smithsonian, hope to increase the number of collection users by expanding the awareness of their collections to audiences beyond those able to travel to the museums themselves. The Smithsonian and several other museums, also have put a great deal of effort into traveling exhibits of various sorts. These traveling exhibits effectively bring the institution to large groups of people who might not otherwise have the opportunity to visit. Still, a traveling exhibit cannot reach everyone; but, the Internet can make an exhibit available to every single home, library, or school quickly, cheaply, and simultaneously. This approach was pioneered by the University of California's Museum of Paleontology (UCMP, 2002b), which was one of the first 25 sites on the World Wide Web in the early 1990s. The Library of Congress (2002), through its American Memory Project, also has been extremely successful in sharing its collections with the nation by these means.

The application of *geoinformatics* may facilitate geoscience data outreach and discovery (see Sidebars 4-7 and 4-8). In such a scenario, all metadata would be digital and accessible over the Internet. Each sample could be located by its geographic coordinates, and metadata would record the circumstances under which the sample was collected, and provide quality control. Such a system would require standardized formats for data archiving, software support, and data-mining tools, and a knowledgeable end-user community. The Kansas Geological Survey's (2001) Data Resources Library geoinformatics systems provide geoscience data over the Internet. Other state geological surveys that have sophisticated data retrieval capabilities over the Internet include Iowa (GEOSAM online database; Iowa Department of Natural Resources, 2002) and North Dakota (fossil, and soon core, database; North Dakota State Geological Survey, 2001).

SUMMARY

There are multiple, necessary steps in preserving and making accessible geoscience data and collections. Digital catalogs available over the Internet are critical to successful

> **SIDEBAR 4-8**
> **Smithsonian's Research and Collections Information System**
>
> The Smithsonian's National Museum of Natural History (NMNH) is creating a Research and Collections Information System that approaches an informatics-based system. The intention is to accomplish three main goals: allow collections management to better track the disposition of specimens acquired, loaned, borrowed, or disposed of, and their location; enable online access to all digital specimen data for the benefit of museum research, collections, and public programs staff, scientists, and the general public worldwide; and to facilitate participation in national and international informatics initiatives. With a suite of software applications, which are used internationally, the staff has begun to implement the systems in a number of science departments. The software was chosen for its stability, ability to scale, flexibility for diverse NMNH disciplines, interoperability with other systems via conformance to international standards, and ability to customize. An estimated 40 million to 50 million records will adequately represent NMNH specimens at a cost of $55 million to $75 million over the next few years. Currently, funds for data entry are limited, so Smithsonian staff are exploring options for obtaining the needed amount (Ross Simons, Smithsonian Institution, personal communication, 2002).
>
> SOURCE: Input during committee site visit to the Smithsonian Institution, April 2001.

management and use of geoscience data and collections. The existence of such catalogs generates multiple benefits—from enhanced use of the collections, to time and money users save in finding material, to improved ability to plan for financial and staffing needs for the collections. The current extent of cataloging in the United States is limited, however, and is the single greatest inhibitor of effective geoscience data and collections use. The backlog of cataloging in many institutions constitutes a significant burden in itself, and overloaded staff would benefit from digital submission of information about newly acquired geoscience data and collections.

5

Regional Centers: A Model for the Future

In this chapter, the committee highlights some examples of partnerships and consortia that have been successful in preserving and making accessible geoscience data and collections. The committee then weighs the pros and cons of various repository arrangements, including the roles of the public and private sector, and puts forth a model for dealing with the growing volume of valuable geoscience data and collections that may be lost in the immediate future. The model is followed by the committee's best estimate of costs involved.[1] The model presented here is one the committee believes will offer maximum likelihood of blending economies of scale (i.e., large enough to house enough geoscience data and collections to make visiting them worthwhile) with regional interests that foster partnerships and consortia across a variety of scales. This chapter also presents a complementary strategy for the federal government, recognizing that the federal government should be taking similar steps to alleviate its own space problems in parallel with the plan the committee has outlined for the non-federal sector. The overall strategy for managing geoscience data and collections in the United States is rounded out at the end of the chapter by a discussion of incentives that could promote preservation.

PARTNERSHIPS AND CONSORTIA

Partnerships and consortia work when all participants benefit by achieving common goals. Organizations establish partnerships for varied reasons, but most often to conserve, stretch, or leverage the resources of time, space, and personnel. Organizations share responsibility, gain efficiency and economy, contribute complementary skills or unique attributes, spread risk, and derive benefit through these cooperative agreements. Partnerships also diminish competition for limited resources, thus avoiding fragmented efforts and duplicated data. In establishing partnerships or consortia, memoranda of understanding (MOUs) document the ground rules of the relationship and, among federal and state agencies, facilitate the transfer of funds (see for example Sidebar 3-5).

Two such partnerships are noteworthy for their successful organization and management, and have been described in chapter 3. The Ocean Drilling Program is managed by the non-profit Joint Oceanographic Institutions, Inc. (JOI), a consortium of 14 oceanographic institutions in the United States (see Sidebar 3-3 and Figure 4-1). Through a series of focused scientific and technical advisory boards, procedures have been established for all aspects of acquisition, curation, sampling, access, and publication of the wealth of marine geologic data collected. The committee was impressed with the distributed core-repository model of ODP and the oversight the scientific community provided.

Administrators of several well-managed geoscience collections have learned that user-defined groups or committees of interest are particularly qualified to define policy concerning access and sampling of specific data types, and to advise on accession and deaccession of materials. This is exemplified at the National Ice Core Laboratory (NICL), which represents a successful partnership jointly sponsored by the NSF and the USGS, and managed with additional NSF funds through the University of New Hampshire (see Sidebar 2-11).

On the personnel side, partnerships with volunteers greatly enhance the capability of institutions operating with limited financial resources. Almost all museums, for example, depend heavily on volunteers to staff basic operations, including curation of collections. Training these volunteers, however, can be time-consuming, so many museums

---

[1] It was beyond the scope of the committee's task to perform a full market analysis of these options. The cost estimates in this chapter are intended to give an overall impression of the types and approximate costs involved in maintaining geoscience data and collections and making them publicly available.

have regularly scheduled training programs that qualify volunteers to work in their collections. One of the most extensive such programs is operated at the Denver Museum of Nature and Science (formerly the Denver Museum of Natural History). Since 1990, the Museum's Certification in Paleontology program has trained more than 150 people in fossil curation and preparation. Many of the program's graduates become expert in various phases of paleontological collections work (see Johnson, 2001).

The foremost responsibility of any curation facility is maintaining the collection. Additionally, a curation facility should demonstrate experience in managing large volumes of data or samples through their various stages of curation. It should have established credibility as a stable institution with a track record for providing the user with the requested service. Finally, it should possess the financial means for long-term survival, adequate upkeep of and access to the collections, and future expansion. The Bureau of Economic Geology at the University of Texas, and C&M Storage, Inc. are examples of two organizations that seemingly fulfill these criteria in the public and the private sector, respectively (see Sidebars 3-4 and 3-1, respectively).

REPOSITORY ALTERNATIVES: IS ONE TOO FEW? ARE 100 TOO MANY?

Options

In this section, we discuss the pros and cons of some of the alternative options for repository size and operation. Table 5-1 summarizes some of the more relevant factors of repository scale. These features are discussed in more detail below, using comparison of trade-offs between features as a means of assessment.

The committee used one basic assumption in considering the pros and cons of repository scale. They assumed that any sort of consortium (i.e., any combination of government, private, and public) would be better than any single entity alone. Consortia provide the kinds of partnership strengths that are missing from a private-only or government-only approach. Consortia also allow broader user-group participation. In addition, consortia allow more flexible funding options and provide the ability to leverage funds from several sectors rather than relying on just one.

Space versus Economy of Scale

The issues of space and economy of scale are, in the committee's opinion, two of the most critical issues to consider when assessing repository size. Certainly large repositories benefit from an economy of scale—that is, fewer administrative costs per item are incurred to oversee and operate a large repository than several smaller repositories that would hold the same amount of material. At some point, however, the sheer magnitude of geoscience data and collections can work against the economy of a single facility. Costs can run quite high for larger and larger buildings, or more and more smaller, interlinked buildings simply because of the architectural constraints of increasing building size and the land-access constraints of increasing numbers of smaller buildings on the same site. The amount of geoscience data and collections that would have to be amassed in a single place is enormous. Even by conservative estimates (see Table 2-1), the amount currently available for donation

TABLE 5-1 Qualitative Assessment of Repository Options

Scale	Access[a]	Clientele Support[b]	Data Variety[c]	Economy of Scale[d]	Space[e]	Time[f]
Single, national	C	C	A	A	F	A
Multiple (several dozen to 100+), sub-regional	A	A	C	F	A	C
Multiple, regional	**B**	**A**	**A**	**B**	**A**	**B**

A = Feature is considered a positive.
B = Feature is considered to be more positive than negative.
C = Feature is considered to be more negative than positive.
F = Feature is considered a negative.

[a]As used here, access represents assumptions regarding ease and cost of travel to a location for users.
[b]Support by clientele includes user-community participation and support for a given facility.
[c]Data variety assumes that different types of geoscience data and collections held in the same place would be beneficial, and that larger facilities hold more and more kinds of data and collections.
[d]Economy of scale assumes that a single larger facility is more economical to operate per volume than a number of smaller facilities that would be required to manage an equivalent volume of geoscience data and collections.
[e]Space is the space required to house geoscience data and collections.
[f]Time is represented in two aspects. First, the time it takes to locate facilities that hold geoscience data and collections of direct interest to the project. Second, the time it takes to visit the facility or facilities.

would require a single facility at least 20 times the size of the USGS Core Research Center in Lakewood (Sidebar 3-2). A facility this size could resolve only an immediate problem (current geoscience data and collections) and would have to be twice its original size in fewer than 15 years assuming even a modest holdings growth rate of 5 percent annually. In other words, it does not take long for a single facility to transcend from utilitarian to onerous.

In contrast, numerous (several dozen to 100 or more) small, sub-regional facilities do not suffer from the space problem of a single, large facility. If space is an issue, another facility can be built or used somewhere else in the region. But what happens if several of these facilities are scattered across a wide area? Although space is much less an issue in this instance, the economy of scale is lost—especially with regard to administrative and managerial costs. Admittedly, some of these costs can be offset by leveraging time or money in local partnerships (with universities, state geological surveys, local geological societies, etc.), but the effectiveness of these arrangements runs the gamut from very effective to completely ineffective, depending on the commitment of the individuals and organizations involved. The likelihood of ineffective support for the long term increases with increasing volunteerism and part-time administrative duties. Long-term support commonly is mediated on the basis of an individual or a few individuals who believe that the use of their time (often with little or no financial reward) is a worthwhile price to pay for conservation of the materials. There is no guarantee that such individual or institutional commitments will continue into the future.

The committee believes that multiple, regional facilities would accomplish economy of scale without spreading resources too thinly. Multiple, regional facilities also require both an institutional and individual commitment on a scale that precludes operation with only volunteers and part-time staff. In addition, multiple, regional facilities would allow common-sense, user-defined distribution of repositories without the requirement that all materials reside in a single place, irrespective of interest or volume, as might be the case for a single, national facility. The committee assumes that multiple, regional repositories would evolve over time to serve smaller or larger regions, depending on the interests of the user communities, volume of geoscience data and collections, and support within the region for multiple facilities. Consequently, one of the key features of the multiple, regional facilities approach advocated here is that it allows participation by a larger number of existing facilities than would a single, national repository. Many of these existing facilities already have a fine reputation among their constituent communities and would add both credibility and economy.

Clientele Support and Data Variety

Clientele support is a critical component of the success of any repository of geoscience data and collections and can be considered from two perspectives: clientele's support for the repository, and the facility's support for clientele. In general, local clientele will directly support local facilities because of easy access, familiarity, and service. A single, large national repository is less likely to garner this type of support unless the holdings are relatively small, yet complete and highly specialized for a specialized community with broad geographic distribution (e.g., Sidebar 2-11). In this regard, support by clientele is somewhat analogous to local library support versus support for the Library of Congress. The former relies much more heavily on the user community (bottom-up approach), whereas the latter relies much more heavily on government funds (top-down approach).

A repository's support for clientele typically is relationship-based. In other words, users of a repository become familiar with the personnel, equipment, and holdings, and thus feel that the repository and its staff support them. Most users believe that smaller facilities allow them the opportunity to develop these sorts of personal relationships far more easily than do larger facilities. Smaller facilities work well for users with specific areas of interest or particular topics of focus, especially if the user can visit only a few of these facilities.

The trade-off here is in data variety. In general, larger facilities have more (both numbers and variety) and broader (geographic, stratigraphic, and variety) types of geoscience data and collections than do smaller facilities. The Smithsonian Institution has a wide variety of geoscience data and collections, along with a wide variety of biological, anthropological, and other data and collections. Such a venue offers much potential exchange of ideas and interdisciplinary collaboration. These are, of course, generalities and not absolutes—some current single, national repositories are highly focused, while some sub-regional repositories are quite varied in their holdings (e.g., NICL and some state geological surveys, respectively).

Regional facilities would offer the feeling of local support that clientele like, with the variety and size of holdings. Moreover, regional facilities may provide a mechanism to divide holdings logically among already existing geoscience data and collections repositories with already established specializations in the area. For instance, if there were a Western Canada Region, one logically could make the case for repositing oil company cores and cuttings in Calgary, Alberta, mining cores at a facility in British Columbia, and paleontology collections at Drumheller, Alberta.[2]

---

[2]Other facilities in a so-called Western Canada Region may have equal claim to any of these specialty areas. This example is presented as an illustration only and is not meant to imply any recommendation of any type by the committee.

Access versus Time

As used here, access represents assumptions regarding users' ease and cost of travel to a location. For instance, the majority of potential users of geoscience data and collections likely would have to travel long distances to use a single national repository. Distributing many repositories through numerous small sub-regions would lessen that burden. Similar arguments apply to the financial burden of transporting new acquisitions to (or borrowing materials from) a national repository versus regional repositories.

Time is represented in two aspects. First, it is represented in the time it takes to locate facilities that hold geoscience data and collections of direct interest to the project. Fewer facilities mean less time spent finding the one(s) that hold relevant geoscience data and collections, whereas more facilities typically mean more time required to track down the appropriate geoscience data and collections.[3] Second, time is represented in how long it takes to visit the facility or facilities. In some instances, all relevant material will be held at a local facility. However, the likelihood of this decreases with increasing number of facilities within any given region.

The committee believes that regional centers represent the best balance of the time versus access issues. Regional centers allow greatest ease of access with most efficient use of time, both to locate holdings and to use the materials once onsite.

Private, Public, or a Combination of the Two?

The committee visited both private and public geoscience data and repositories (see list in Preface). Private repositories generally fall into two categories: those that hold physical items and those that hold electronic data. Private repositories that hold physical items (cores, cuttings, paper and fiche well logs) typically either hold them in propriety for a fee (e.g., C&M Storage Inc.), or amass paper, fiche, and other physical records that already are public, but have not otherwise been assembled conveniently into one place (e.g., Denver Earth Research Library), for which users typically pay either a fee or monthly membership/subscription for use. Private repositories that hold electronic data typically either acquire publicly available data in non-electronic form, then convert them to electronic data, or purchase or otherwise acquire electronic data, which they own. These companies also allow access to data either on a fee or subscription basis. Privately operated facilities in the latter two categories add value to the geoscience data and collections by collecting them in one place (that is, doing all the "leg-work" for a potential user) or by editing and processing the data (i.e.,

adding value through information and interpretation) or some combination of the two. In other words, they make money by adding value and selling access to this added value. The versions of these entities handling non-digital (paper-based) data (e.g., DERL, see Sidebar 3-9) typically are operated as non-profit entities.

Those entities that hold physical data in propriety are acting as contractors for companies (mostly oil companies) that want access to their geoscience data and collections, but do not have the facilities to keep them onsite. In this situation, access to the geoscience data and collections cannot be purchased directly from the holding company, which is acting as a subcontractor for the actual owners of the material. Companies that operate in this mode make money by retaining confidentiality and privacy of the geoscience data and collections, along with added services such as delivery, pick-up, shipping, and repackaging of the samples for their clients.

In contrast to private companies, public entities typically operate with public subsidies (local, state, or federal) and make their holdings publicly available. Such institutions commonly hold geoscience data and collections as part of their charter (e.g., the BEG at the University of Texas, see Sidebar 3-4), in association with some other function, such as a museum or institution of higher learning, or some combination of both. Rarely do these public institutions receive additional funds beyond their annual allowances for anything other than beyond-normal services (e.g., cutting a core for examination by an outside visitor).

Private entities have the advantage of being able to charge for services and recover costs for all operational expenses and, in some cases, even for profit. However, private entities are subject to the ups and downs of a market-driven economy, which can have a destabilizing effect. Public entities have the advantage of being more stable in the long run than most private ones, but they have the disadvantage of usually operating with minimal support and few options to recover costs, much less make additional money for data and collections support. The committee believes that a combination of public and private entities, operating as a consortium, offers the best combination of stability and fiscal opportunity.

THE REGIONAL CENTERS CONCEPT

The committee found that successful preservation and research centers generally served relatively focused communities of interest, most often (with a few notable exceptions) geographically defined (for example, the Bureau of Economic Geology at the University of Texas, Sidebar 3-4). Such regional centers are large enough to achieve economies of scale, but small enough to encourage local interest and support. Distributing the centers would permit sponsors to nurture regional networks of dedicated volunteers, data and collections donors, and financial benefactors. As the focal point of a regional consortium, each center would draw upon existing expertise and infrastructure, such as state geo-

---

[3]This assumption is not met when all the repositories in an area share data and are collectively searchable at one time using a metadata search tool.

logical surveys, museums, universities, and for-profit (private) enterprises.[4] Each center also would encourage adoption of uniform standards, coordinate outreach efforts, manage effective use of existing space and addition of new facilities, and facilitate cooperative projects and sharing of resources and expertise among the centers and consortia members.

Because regional centers serve a national interest, they warrant several forms of federal support. A key component would be to facilitate establishment of new centers on a competitive, cost-sharing basis. The committee's investigations indicate that it is generally easier to acquire support to help establish the centers than to operate them on a long-term basis. Consequently, providing federal funds for those activities for which it is most difficult to get continuing sponsor support (e.g., cataloging, curation, operations, maintenance) is a critical component of the support needed for establishment and long-term operation of the regional centers. The innovative model used in the Shell Oil donation to the Bureau of Economic Geology, University of Texas (see Sidebar 2-2), could greatly reduce the level of continuing federal support required. In this instance, the donor contributed a sizable endowment to cover recurring operating expenses. A DOE grant permitted the endowment to grow to the point where its proceeds will cover projected costs in perpetuity. The costs of cataloging, packaging, and delivery of newly collected materials, consistent with established data standards and best curation practices, as part of the collection acquisition, would be necessary components of any successful regional center model.

Regional centers should be able to set appropriate charges, consistent with community interests and ability to pay, for selected services (if they wish to do so). In addition to reimbursement for the services, these charges could be used for long-term growth and expansion of capabilities and facilities. By way of illustration, a service charge could be applied when busy clients would rather pay someone else to examine a set of samples and compile a report, rather than spend time doing it themselves. Such selected service charges should be market-based and revenues should return directly to the center's budget.[5]

Several institutions supporting oil and **gas** exploration cover substantial portions of annual expenses by subscriptions and charges for specific services. While the regional centers may offset some of their operational costs with these fees, they are unlikely to become entirely self-sufficient. Furthermore, some repository staff expressed concern that user fees would discourage use. Consequently, to hold some of these user fees to a minimum, all the centers are expected to require some level of government funding, although the amount may vary. For example, centers with less prosperous user communities (such as those in education and research) or centers with fewer users may require a higher level of support.

The committee became convinced that a uniting theme of the successful centers is direct involvement of an external science-advisory board. Such boards, composed primarily of users of the facility, help establish priorities for geoscience data and collections acquisition, in addition to facilitating maximum utilization of the data and collections by the widest possible range of clientele.[6] Given the success of analogous centers with NSF input, and NSF's support of external science-advisory boards, the committee was convinced that NSF is an appropriate federal agency to award federal funds for the proposed regional centers. Following the NSF model, such funds would be awarded on a competitive basis, with preference given to consortia with an active, external science-advisory board and broad participation from government, academia, and the private sector.

Three such centers (one each in the Gulf Coast, Rocky Mountain, and Pacific Coast regions) are needed immediately. These three areas currently have the most critical need for preserving geoscience data and collections. First, the volume of physical data in these areas is overwhelming because of the long history of resource extraction (see for example Table 5-2). In addition, many of the risk factors outlined in Table 2-4 occur in these regions, including shifting priorities of those holding data, and industry mergers. These regions also contain a wide range of clientele, both private and public. Finally, these regions contain many examples of good practices and successes on which regional consortia can be built.

The committee does not intend to limit regional centers to these three regions, nor does it intend to limit any of these regions to one center only. However, one center established in each of these three regions would provide immediate, critically needed relief for the growing problem of geoscience data and collections loss.

Cataloging will be an enormous but essential task for all regional centers. It was apparent to the committee from site visits to the National Museum of Natural History, Bureau of Economic Geology at the University of Texas, and USGS Core Research Center that cataloging is extremely important, yet time-consuming, and that it requires a great amount of staff effort. In addition, adequate computer software and hardware required for online availability of the catalog information can be costly to the average institution that maintains geoscience data and collections.

---

[4] The committee anticipates that existing repositories likely would participate in regional consortia though additions to existing facilities. Participation of new entities is not precluded, however.

[5] In the committee's opinion, government policies that direct revenues to a general fund discourage local initiative and responsiveness to user needs.

[6] The science-advisory board, while predominantly composed of members of the user community, would benefit from expertise in database management issues, including digital cataloging.

TABLE 5-2 Percentage of Total U.S. Oil Production, 1945–1975 and 1976–2000, as a Proxy for Volume of Geoscience Data and Collections in the Gulf Coast, Pacific Coast, and Rocky Mountain Regions

Region and State	Average percentage of U.S. Crude Oil Production	
	1945–1975	1975–2000
Gulf		
Alabama	0.19	0.64
Arkansas	1.14	0.48
Kansas	4.08	1.96
Mississippi	1.77	1.02
Louisiana	16.50	17.42
Oklahoma	7.38	4.32
Texas	39.14	27.91
Pacific		
Alaska	0.67	19.40
California	13.17	13.02
Rocky Mountain		
Colorado	1.30	1.06
Montana	0.89	0.83
New Mexico	3.34	2.66
Utah	0.64	0.95
Wyoming	4.03	3.70
Total percentage	**94.05**	**95.44**

SOURCE: DeGolyer and MacNaughton, 2001.

Much must be accomplished quickly in the area of cataloging to maximize the effects of newly available information about existing geoscience data and collections. The magnitude of this problem cannot be overstated—the geoscience community, in particular, and the nation in general has inadequate information about the amount and condition of current geoscience data and collections (see chapter 4). Cataloging is the first step to assessing the quality of documentation of (and hence whether to keep or discard) geoscience data and collections currently in the nation's museums, institutions, agencies, and repositories.

Cataloging of other collections also should be encouraged—for those entities wishing to join consortia, as well as simply to increase knowledge and use of other collections outside the consortia. The committee believes that awards for cataloging should be distributed on a competitive basis, using general priorities for preservation as outlined in Table 2-5 as a guide to need. The Institute of Museum and Library Services has, since 1996, distributed on a competitive basis federal funds for improving access to information about collections at museums and libraries (Sidebar 4-4), and could be an appropriate administrator of federal funds for cataloging geoscience data and collections.

Costs for Regional Centers and Cataloging

The realities of preserving geoscience data and collections implies incurring costs if the issues outlined in this report are to be addressed—geoscience data and collections occupy space, they cost money to curate and manage, and they are critical sources of information. Although not specifically charged with providing cost estimates to implement the broader strategies proposed herein, the committee believes that it is in the best interests of those who might want to follow the recommendations to have some general estimates for the minimal costs required to do this effectively. The tables below show the committee's reasoning and underlying assumptions about potential costs of one implementation strategy.

Given the estimate that sufficient data are at risk of loss to fill more than 20 times the volume of the USGS Core Research Center in Lakewood, Colorado, the committee was convinced that solving the problem requires new repository space rather than solely increasing the efficiency with which existing space is utilized.[7] Certainly, some portion of initial funding could (where practical) support more efficient utilization of existing space. Additionally, more than one new repository might be needed within a single regional consortium. However, in the interests of simplicity, the following calculations assess the cost of a single, new facility in each region. Obviously costs can vary considerably, but the ranges provided should cover most of the possibilities irrespective of the specific circumstances under which they might be applied. Although offered as guidance and estimates only, the committee is united in its belief that this is the appropriate range of costs needed to address the critical task of preserving important geoscience data and collections.

The committee foresees start-up costs of $35 million to $50 million per center.[8] The lower end of the range assumes a center of similar magnitude to the BEG (see Sidebar 3-4), while the higher end is for a facility that is 50 percent larger than the BEG. The costs could be spread over several years for initial investments and rescue of highest-priority materials threatened with imminent loss (Table 5-3). The federal government has an important stake in the establishment of these repositories. Therefore it should provide for a major share of their costs. The committee's cost estimates assume

---

[7] By "new repository space," the committee recognizes that existing repositories likely would participate in regional consortia though additions to existing facilities. "New repository space" neither implies nor precludes new buildings in new places operated by completely different organizations, it simply acknowledges the critical lack of space in existing facilities without new construction.

[8] The cost estimates are based on discussions with Ronald Broadhead, New Mexico Bureau of Geology and Mineral Resources, George Bush and Douglas Ratcliff, Bureau of Economic Geology, Robert Shafer, C&M Storage, Inc., and Guenter Wellman, Alberta Core Research Centre, Canada. All cost estimates are in 2001 dollars.

TABLE 5-3 Estimated Cost Range to Establish a Regional Center[a]

	Low	High
Volume and Space Assumptions:		
Core	600,000 boxes	900,000 boxes
Cuttings	800,000 boxes	1,200,000 boxes
Immediate space use	60,000 sq. ft.	90,000 sq. ft.
New acquisitions	60,000 sq. ft.	90,000 sq. ft.
Total building space	120,000 sq. ft.	180,000 sq. ft.
Capital Costs:		
Construction costs[b]	$ 7,000,000	$ 9,000,000
Land acquisition[c]	$ 50,000	$ 75,000
Shelving[d]	$ 2,000,000	$ 3,000,000
Equipment; furniture[e]	$ 1,000,000	$ 1,500,000
Total Capital Costs	$10,050,000	$13,575,000
Moving Costs:		
Packing; inventorying[f]	$21,000,000	$31,500,000
Shipping[g]	$ 1,400,000	$ 2,100,000
Total Moving Costs	$22,400,000	$33,600,000
Other Set-up Costs:	$ 2,000,000	$ 3,000,000
Total Start-up Costs:	$34,450,000	$50,175,000

[a]Costs based on information provided by Ronald Broadhead, New Mexico Bureau of Geology and Mineral Resources; George Bush and Douglas Ratcliff, Bureau of Economic Geology; Jimmy Denton, BP Amoco, Tulsa, Oklahoma; Robert Shafer, C&M Storage Inc.; and Guenter Wellmann, Core Research Centre, Alberta, Canada 2001.
[b]Construction costs based on $50/ft^2.
[c]Land acquisition costs probably vary more with urban versus rural settings than with regions.
[d]Shelving costs estimated at 1.4 boxes of material per dollar of shelving.
[e]Includes lab equipment, office equipment, core-retrieval equipment.
[f]Includes labor, cataloging, container costs, other supplies related to packing and labeling.
[g]Estimated at $1 per box of material shipped 500 miles.

that federal funding will be primarily for constructing storage for physical specimens; however, some centers may require relatively more funds for geoscience data and collections rescue. Capital improvements would be required from time to time to accommodate additional holdings.

Start-up Costs

Capital Expenses

Facilities are a significant part of the cost to establish regional geoscience research centers, and their costs vary widely, due in part to differences in land costs and, importantly, to differences in climate and the degree of protection required for materials to be preserved, and to the support services available for users. Labor (construction) costs also differ widely and contribute substantially to capital expense variability. For simplicity, the committee bases its estimates on building anew, and recognizes that costs could be less if a center were to build off existing infrastructure.

Moving Expenses (Rescuing Physical Specimens)

Rescue of physical specimens is costly and labor intensive. Holdings must be inventoried, culled (where necessary), and cataloged—all labor-intensive efforts. Boxes must be packed onto pallets, loaded into shipping containers, shipped, unloaded, unpacked, and shelved. Cataloging costs vary with the unit of material to be cataloged—one well may yield hundreds of boxes of core, but for other materials, one box may contain many specimens that require individual catalog entries.

Costs for inventorying and physical handling vary widely, from less than $10 per box to as much as $30 per box (Robert Shafer, C&M Storage Inc., Schulenberg, Texas, personal communication, 2001; Susan Longacre, ChevronTexaco, Houston, Texas, personal communication, 2001). In general, the volumes of material moved are thousands to several hundred thousands of boxes, and the $10-per-box estimate is more appropriate. Shipping costs also vary; however, the committee estimates a cost of about $1 per box for 500 miles. Fragile and sensitive materials, or those that require special handling (e.g., samples that must be refrigerated to preserve delicate organic compounds) likely would cost more than the estimates used by the committee.

Surprisingly, many of the same expenses are incurred to dispose of geoscience data and collections. Disposal does not require cataloging and usually does not involve special handling, but disposal does require inventorying, testing for hazardous content, packing, loading, shipping, and unloading, in addition to disposal fees.

Recurring Costs

Operating Costs

Recurring costs of research center operation include facilities operations and maintenance, data-center operations, and support for the wide range of acquisition, curation, preparation, outreach, user support, and active research operations described above. The committee estimates that operating budgets of $3 million to $5 million per year are needed for each center (see Table 5-4). Because of the federal government's important stake in the effective operation of these centers, it should provide for a major share of these costs. Moreover, federal support would provide effective leverage of support from other sectors. These remaining recurring costs should come from local consortia, service charges, and other sources.

TABLE 5-4 Estimated Range of Recurring Costs for Each of the Three Proposed Centers

Recurring Costs (Annual)	Low	High
Staff[a]	$ 500,000	$1,000,000
Facilities[b]	$ 75,000	$ 85,000
Travel; computer center[c]	$ 100,000	$ 150,000
New acquisitions[d]	$2,240,000	$3,360,000
Total per year	**$2,915,000**	**$4,595,000**

[a]Staffing costs are for 4-8 total full-time employees at approximately $125,000/person.

[b]Facilities costs include utilities, operations and maintenance (not insurance or supplies).

[c]Costs include staff travel to evaluate potential donations and partial shipping expenses for loans.

[d]New acquisitions costs include packing, shipping an average of 500 miles, cataloging, necessary re-boxing (SOURCE Robert Shafer, C&M Storage, Inc., personal communication, 2001; Susan Longacre, ChevronTexaco, personal communication, 2001; Jimmy Denton, BP Amoco, personal communication, 2001).

TABLE 5-5 Representative Service Charges

Type of Service	Amount
Subscription access	
Hard-copy holdings	$100-$1,150/year
Digital publications	Vary widely, depending on scope, detail, size of using organization, etc.
Data purchase/reproduction	
Electronic catalogs	$40-$1,000 on CD-ROM
Reports:	
Electronic	Unlimited downloads to CD-ROMs at $1,000 and up
Paper, microfiche	$2/report to $25,000/series
Catalog access	Free to $0.15/screen
Physical access to specimens	
Box retrieval from storage	$1.50-$10/box; no additional charge for some storage services
Facility use for examination	$30-$150/day

SOURCE: Fee schedules for NGDC Marine Geology and Geophysics Center, Alberta Core Research Centre, and the Bureau of Economic Geology Core Research Center, University of Texas.

New Acquisitions

One of the more striking revelations was the inability of those institutions that constructed new geoscience data and collections facilities to anticipate adequately the amount of time required to fill the space. For example, new facilities at the state geological surveys of Kansas, Kentucky, and Ohio, constructed in 1990, 2000, and 2001, respectively, now have <5, 10, and 16 percent space available, respectively (see Table 2-3a). Although acquisitions likely will be sporadic, volume undoubtedly will be high.

Summary of Recurring Costs

Table 5-4 presents the committee's estimate of recurring costs for each of the three proposed centers.

Costs: Independent Cataloging Support

Cataloging is an assumed component of the staff duties at each of the regional centers. It is also a component of the committee's overall strategy for managing geoscience data across the nation. Major advances in cataloging could be achieved if approximately $1 million were available to each of 5 to 10 institutions annually on a competitive basis.

Access Charges

Several of the repositories reviewed by the committee charge for services and in some cases recoup a sizable portion of their operating revenues in this way. Table 5-5 presents a sample of charges at several active institutions.

ADDITIONAL ROLES AND RESPONSIBILITIES OF THE FEDERAL GOVERNMENT

Federal agencies responsible for geoscience data and collections in the United States should lead the way by setting examples of good practices in preservation of and access to geoscience data and collections. Such examples serve to promote the public good, increase the visibility of the federal side in a leadership role, and increase the likelihood of federal partnerships with the private sector.

While it exists, coordination among federal agencies that collect or archive geoscience data and collections could be improved. Such improved coordination would optimize sharing of business practices and consumer use of related data collected by various agencies or establishing priorities across agencies so that limited funds can be used to the best overall effect. Chapters 1 through 4 contain examples that highlight the benefits of strong coordination, and the consequences of poor coordination. Adoption of consistent and good practices, along with a clarification of roles, would, at a minimum, increase efficiencies for federal agencies and the user community, comparable in some respects to the goals of the National Spatial Data Infrastructure (NRC, 1993) and the *Geospatial* One-Stop initiatives.[9] In addition, such collabo-

---

[9]These two initiatives are useful models in several respects. First, they seek to render data from many federal, state, and local agencies both convenient to access and easy to use together. Second, they must address diverse missions, user communities, producer concerns, data definitions, and data formats. Information providers may themselves produce the data, or they may obtain it from external sources. Coordination of U.S. geoscience data and collections will involve all of these issues.

TABLE 5-6 Proposed Roles of a Federal Geoscience Data and Collections Coordinating Committee and Federal External Science-Advisory Boards

Roles of the Federal Geoscience Data and Collections Coordinating Committee	Roles of the Federal External Science-Advisory Boards
• Determine how to coordinate and streamline federal efforts in preservation of, access to, and use of geoscience data and collections • Monitor conformance to agreed-upon practices • Monitor and facilitate progress of cataloging efforts across the federal government • Monitor implementation of electronic reporting for all exploration, exploitation, and research reports currently submitted to the federal government • Facilitate and coordinate Internet access to all federal geoscience data	• Advise on priorities for federal holdings, with respect to preservation, cataloging, and access across and within federal and quasi-federal agencies • Advise on establishing consistent practices across agencies with respect to preservation of and access to geoscience data and collections acquired from public lands or using federal funds • Coordinate with science advisory boards of regional consortia

ration would render the whole of government holdings more complete, enhance the value of individual components, and permit a significantly (and, eventually, measurable) increased benefit to diverse communities.

A federal geoscience data and collections coordinating committee would address the problem. Such a committee could be established and funded through the Office of Management and Budget, and would oversee coordination and increased efficiency among a range of federal agencies. This federal geoscience data and collections coordination committee should be broad-based, reaching between and within all federal and quasi-federal agencies that are involved in geoscience research or geoscience data and collections acquisition. The committee's charge should focus on coordination of federal agencies' roles with regard to geoscience data and collections preservation, access, and use.

Parallel with its coordination and streamlining activities, the federal geoscience coordinating committee should establish federal external science advisory boards to advise on priorities for federal holdings, with respect to preservation, cataloging, and access across and within federal and quasi-federal agencies. Previous NRC reports (e.g., NRC, 2001) already have noted the value for federal agencies of having direct external community involvement and advice in order to help set internal priorities for funding, monitoring, and research efforts. Examples of existing federal external science-advisory boards that deal with collections are those within the operating structures of the National Ice Core Laboratory (coordinated jointly by the USGS and NSF; see Sidebar 2-11) and the Smithsonian Institution (Smithsonian Institution, 2001).

The federal external science-advisory boards would focus on holdings within the federal government, but would coordinate with the science advisory boards of the regional geoscience data and collection centers. The federal external science-advisory boards, which could be discipline-based, would advise on establishment of consistent practices across agencies with respect to preservation of and access to geoscience data and collections acquired from public lands or using federal funds. In addition, the federal external science-advisory boards would advise on what geoscience data and collections should fall within the purview of various federal agencies. Monitoring of conformance to agreed-upon practices, as a question of how, rather than what, would reside within the charge of the federal geoscience data and collections coordinating committee. Table 5-6 summarizes the proposed roles of the federal coordinating committee and the external science advisory boards.

The federal geoscience data and collections coordinating committee would have other responsibilities related to how the federal effort should be streamlined, coordinated, and improved. One such responsibility would be monitoring implementation of electronic reporting for all exploration, exploitation, and research reports currently submitted to the federal government. The committee believes that electronic reporting is a necessary step to minimize the burden of cataloging newly collected geologic samples, while maximizing their potential use. As noted earlier, cataloging, providing electronic access to, and advertising the availability of existing geoscience data and collections will be immensely challenging. The National Science Foundation's FastLane system is an example of agency effort to coordinate reporting practices and make those results available to the community of geoscience professionals as well as to other potential users. Examples of other programs of electronic reporting exist at the provincial level in Canada and Australia, and in the state of Wyoming.

The cataloging effort recommended for non-federal institutional holdings is of equal importance for future use of federal geoscience data and collections holdings. Therefore, the federal geoscience data and collections coordinating committee should monitor and facilitate progress of cataloging efforts across the federal government. Here, the federal geoscience data and collections coordinating committee would work closely with the federal external science-advisory boards to determine which cataloging efforts warrant the highest priorities (using the general priorities identified in Table 2-5, supplemented with data-specific advice on potential future applications). In addition, the federal geoscience data and collections coordinating committee should facilitate and coordinate Internet access to all federal geoscience data. This would include (but not be limited to) reports and catalogs of holdings, location and availability of similar geoscience data and collections, and contact information (where appropriate) for onsite use of geoscience data and collections. Success of this effort will be enhanced by coordinated adoption of digital data standards to improve interoperability of interagency information.

Regular review of the roles of the National Science Foundation and Institute of Museum and Library Services as distributors of funds for non-federal cataloging and repository efforts is essential. If existing external review mechanisms (e.g., committees of visitors; external steering committees) are inadequate for this task, new ones should be devised.

Federal Involvement in Regional Consortia

In addition to coordination between federal external science-advisory boards and those of the regional centers, the committee anticipates that federal agencies would participate (where appropriate) as partners in the regional consortia proposed earlier in this chapter. With large volumes of potentially useful geoscience data and collections at risk within federal government agencies (see for example chapter 2, and Table 2-3b), new federal geoscience repositories also are warranted. Start-up and recurring costs for such repositories would parallel costs outlined in Tables 5-3 and 5-4, respectively. The committee envisages funding for federal and non-federal entities converging within the regional consortia in instances of federal participation in such consortia. For example, arrangements already exist between state and federal agencies in Alaska (Sidebar 3-5) and Colorado (Sidebar 3-2). Priorities for federal agency support should follow closely those recommended for the regional centers: need for such a repository within the agency; broad or active involvement within and among various federal geoscience agencies (e.g., BLM, DOE, EPA, NASA, NOAA, NSF, USACE, USGS, USNM); and active participation of federal external science-advisory boards.

INCENTIVES

Incentives for preservation of geoscience data and collections would encourage preservation efforts. Such incentives would encourage private donations of geoscience data and collections, by providing credit for shipping costs and fundamental recognition that fossils, rock, sediment, and ice are unique and have donation value (see chapter 2). When such data and collections are used to enhance recovery of resources, federal support for these incentives has the potential to pay for itself many times over (DOE, 2002). An incentive for the research community is to require that geoscience data and collections amassed during the course of federally funded research (funded by agencies such as DOD, DOE, EPA, NASA, NSF, USGS, USNRC) be appropriately archived and cataloged and made accessible to the public (see for example USGCRP, 1991). Federal support for research should be, in general, contingent upon the public availability of these geoscience data and collections within a reasonable time.

The geoscience community itself must take more responsibility for preservation and use of geoscience data and collections. Although the importance of these data for research and interpretation are broadly accepted, adequate curation and long-term care for them take time, are comparatively unrewarded, and consequently fall through the cracks. The geoscience community must do more than just acknowledge the importance of geoscience data and collections—it should establish incentives, rewards, and requirements for their care and accessibility. The geoscience community should adopt standards for citation of geoscience data and collections used in scientific and other publications. Citation histories lend enhanced credibility and importance to well-organized, often-used data and collections. In addition, institutions and professional societies should establish (where appropriate) awards and other forms of recognition for outstanding contributors to the preservation and accessibility of geoscience data and collections.

6

Challenges and Solutions

The challenges the committee encountered during its investigation into the issues surrounding preservation of geoscience data and collections are a microcosm of those facing the geoscience community and the nation. The conclusions and recommendations the committee reached address these challenges.

One of the main challenges the committee faced was assessing the volume, quality, and location of the nation's geoscience data and collections.

- How much and how many geoscience data and collections currently are preserved and available?
- How much and how many are preserved, but unavailable for various reasons (proprietary reasons, inappropriate storage, lack of knowledge of their existence)?
- Who has geoscience data and collections now?
- What is the current condition of their holdings?
- How much room remains to preserve those that should be retained and to make them accessible?

The committee found information of this type in short supply, but was able to assess the general magnitude of the problem by assembling information where known (Tables 2-1, and 2-2).

A second challenge the committee encountered was determining the nature and effectiveness of the federal-agency effort to preserve and make geoscience data and collections available. Which agencies have responsibility for geoscience data and collections? Over which geoscience data and collections do these agencies have responsibilities? Testimony and input during site visits evinced that while it exists, coordination among federal agencies that collect or archive geoscience data and collections could be improved. Such improved coordination would optimize sharing of business practices and consumer use of related data collected by various agencies or establishing priorities across agencies so that limited funds can be used to the best overall effect. Chapters 1 through 4 contain examples of the benefits of coordinated actions, and the consequences of poor coordination.

Space for present and future geoscience data and collections is a critical challenge. The committee found a critical shortage of space for current geoscience data and collections, let alone those gathered in the future. This challenge itself was not a surprise to the committee—the surprise was the magnitude of the problem that faces the nation and the seeming inability to plan adequately for future space needs. For instance, several state geological surveys that have constructed core libraries within the past 12 years already report 16 percent or less remaining space. Not everything can or should be saved, but the committee was surprised by the reasons given for not preserving endangered geoscience data and collections—mostly related to space and cost as opposed to some of the priorities listed in Table 2-5, for example.

Finally, the committee noted the challenge of rewarding effective curation of geoscience data and collections. Currently, geoscience data and collections are not used to the fullest primarily because of lack of access to information about them, which relates directly to the state of their curation. How can information about geoscience data and collections (i.e., the metadata) be made available more widely? What can be done to encourage effective curation of and access to geoscience data and collections?

The committee recommends the following actions in order to address the challenges outlined above.

1. The committee recommends that priority for rescuing geoscience data and collections be placed on those in danger of being lost. The highest priority for retention and preservation should be directed toward data and collections that are well documented and impossible or extremely difficult to replace.

2. The committee recommends funding cataloging ef-

forts to gather comprehensive information about existing geoscience data and collections. Access to these funds should be on a competitive basis, and preference should be given to institutions with holdings that meet the same priorities as those outlined above for preservation. The committee recommends that this initial catalog funding effort target 5 to 10 institutions each year until the nation's geoscience data and collections are adequately assessed.

3. The committee recommends the establishment of a distributed network of regional geoscience data and collections centers, each with an external science-advisory board. The committee recommends establishing three centers (one each in the Gulf Coast, Rocky Mountain, and Pacific Coast regions). Furthermore, the committee recommends that additional regional centers, as merited, be established over the next 5 to 10 years, and that preference be given to those centers that meet three main criteria: need for such a center in the region (i.e., active clientele, identified collections of high-priority, at-risk data in the region); broad involvement and support among various regional geoscience and other entities (government, academia, and industry); and active participation of an independent, external science-advisory board. The committee recommends that the centers build upon existing expertise and infrastructure, such as state geological surveys, museums, universities, and private enterprises, and that, where practical, more efficient use of existing space be encouraged before expansion. Furthermore, the committee recommends that access to the center-establishment and improvement funds be on a competitive basis.

4. The committee recommends establishing a federal geoscience data and collections coordinating committee to optimize federal coordination. The federal geoscience data and collections coordinating committee should appoint several federal external science-advisory boards to advise on priorities for federal holdings, with respect to preservation, cataloging, and access across and within federal and quasi-federal agencies. In addition, the committee recommends that electronic reporting be implemented as soon as possible, with additional funding as required to accelerate it, to reduce the added burden of cataloging future data and samples. The committee recommends that the federal geoscience data and collections coordinating committee monitor and facilitate progress of cataloging efforts across the federal government.

5. The committee recommends that federal agencies be supported to the same extent as non-federal institutes and consortia with respect to cataloging and repositories, with regular review. The committee recommends that priority for federal agency support follow closely those recommended for the regional centers: need for such a repository in the agency; broad or active involvement within and among various federal geoscience agencies (e.g., BLM, DOE, EPA, NASA, NOAA, NSF, USACE, USGS, USNM); and active participation of federal external science-advisory boards.

6. The committee recommends establishing a combination of federal, state, regional, and local government incentives and requirements for geoscience data and collections, donations, and deposition. Establishing such incentives should be an immediate priority to stem the tide of lost and discarded geoscience data and collections, many of which remain useful.

7. The committee recommends that the geoscience community adopt standards for citation in scientific and other publications of geoscience data and collections used. Institutions and professional societies should establish (where appropriate) awards and other forms of recognition for outstanding contributors to the preservation and accessibility of geoscience data and collections.

Although challenging, the issues the committee encountered during its investigation into preservation of geoscience data and collections are by no means insurmountable. More importantly, there exist both immediate and long-term benefits to preserving appropriate data and collections. Although the immediate benefits often are apparent, the long-term benefits require careful and imaginative evaluation. Examples of such benefits are: enhanced understanding of the nation's natural resources; better assessment of the value and extent of public resource holdings; increased safety of the population through knowledge of potential natural hazards and appropriate engineering to minimize damage; more rapid response to natural hazards emergencies; and better knowledge of the history of life on Earth. The continued loss of potentially useful geoscience data and collections erodes our ability to realize these and other benefits. The recommended steps in this document outline a strategy that will reduce this erosion, but only if acted upon now.

References

AGI [American Geological Institute]. 1994. National Geoscience Data Repository System Feasibility and Assessment Study. Alexandria, VA: AGI, 68 pp.

AGI [American Geological Institute]. 1997. National Directory of Geoscience Data Repositories. Alexandria, VA: AGI, 91 pp.

AGI [American Geological Institute]. 2002a. Georef. Accessed on April 5, 2002 at: http://www.georef.org.

AGI [American Geological Institute]. 2002b. National Geoscience Data Repository System. Accessed March 11, 2002 at: http://www.agiweb.org/agi/NGDRS.

Alley, R. B. 2000. The Two-Mile Time Machine: Ice Cores, Abrupt Climate Changes and Our Future. Princeton, NJ: Princeton University Press, 240 pp.

Allison, M. L. 2001. A geologic detective story. Geotimes October:14-19.

Allmon, W. D. 1994. The value of natural history collections. Curator 37(2):82-89.

Allmon, W. D. 1997. Collections in paleontology. Pp. 155-159 in H. R. Lane, J. Lipps, F. F. Steininger, and W. Ziegler, eds. Paleontology in the 21st Century Workshop. Kleine Senckenbergreihe No. 25. Frankfurt, Germany: Senckenberg Museum.

Allmon, W. D. 2000. Collections in paleontology. Pp. 201-214 in H. R. Lane, F. F. Steininger, R. L. Kaesler, W. Ziegler, and J. Lipps, eds. Fossils and the Future. Paleontology in the 21st Century. Frankfurt, Germany: Senckenberg-Buch.

Allmon, W. D., and M. Lane. 2000. Orphaned and endangered collections in invertebrate paleontology: a call for a national solution. Pp. 43-50 in R. D. White and W. D. Allmon, eds. Guidelines for the Management and Curation of Invertebrate Fossil Collections, special publications vol. 10. Ithaca, NY: Paleontological Society.

American Chemical Society. 1999. ACS Books Reference Style Guidelines. Accessed February 27, 2002 at: http://pubs.acs.org/books/references.shtml.

ARCSS [Arctic System Science Data Coordination Center]. 2002. Greenland Ice Sheet Project 2. Accessed March 11, 2002 at: http://nsidc.org/arcss/projects/gisp2.html.

ARL [Association of Research Libraries]. 2000. ARL Statistics. Accessed January 24, 2002 at: http://www.arl.org/stats/arlstat/index.html.

Australia Department of Industry, Tourism and Resources. 2001. Administration of Mining Law in Western Australia. Accessed March 5, 2002 at: http://www.isr.gov.au/Resources/minerals/guide/leaflet_16.pdf.

Bailey, S. A., T. M. Kenney, and D. R. Schneider. 2001. Microbial enhanced oil recovery: Diverse successful applications of biotechnology in the oil field. Kuala Lumpur, Malaysia: Society of Petroleum Engineers 72129, 8 pp.

Baker, R. 1980. A Primer of Oil Well Drilling (Petroleum Extension Service). Austin: University of Texas, 79 pp.

Baker Hughes. 2001. Drilling Systems: Core Bit Selection Guide. Accessed November 7, 2001 at: http:www.bakerhughes.com/integ/Drilling/coring/bits/arc_bitguide.htm.

Bass, D. M. Jr. 1992. Properties of reservoir rocks. Pp. 26-1 to 26-33 in H. B. Bradley, ed. Petroleum Engineering Handbook. Richardson, TX: Society of Petroleum Engineers.

Bradley, H. B., ed. 1987. Pp. 45-1 to 15, 46-1 to 19, 47-1 to 16, and 29-1 to 53-26 in Petroleum Engineering Handbook. Richardson, TX: Society of Petroleum Engineers.

Bradley, R. S. 1985. Quaternary Paleoclimatology: Methods of Paleoclimatic Reconstruction. London: Unwin Hyman, 472 pp.

CSA [Cambridge Scientific Abstracts]. 2002. Oceanic Abstracts. Accessed on April 5, 2002, at: http://www.csa.com/csa/factsheets/oceanic.shtml.

CSPG [Canadian Society of Petroleum Geologists]. 2001. Preservation of core and drill cuttings survey. Accessed January 28, 2002 at: http://www.cspg.org/Programs_Services/Divisions/Core___Sample/CoreSampleArticle.pdf.

Cooley, G. P., M. B. Harrington, and L. M. Lawrence. 1993. Analysis and Recommendations for Scientific Computing and Collections Information Management of Freestanding Museums of Natural History and Botanic Gardens, MTR 93W0000109. McLean, VA: Mitre, Inc. 2 vols. 1084 pp.

Cranbrook, E. 1997. The scientific value of collections. Pp. 3-10 in J. R. Nudds and C. W. Pettitt, eds. The Value and Valuation of Natural Science Collections. London: The Geological Society.

Davidson, K. 1999. Archaeologists take a dig at online fossil sales. San Francisco Examiner, October 4. p. A-1.

Davis, T. L., and J. Namson. 1994. A balanced cross section of the 1994 Northridge earthquake, Southern California. Nature 372:167-169.

Deal, E. G., D. Wolfgram, and R. B. Berg. 1999. Reno Sales–Charles Meyer–Anaconda memorial collection; a unique sample suite from a world-class mining district. Abstracts with Programs from Annual Meeting in Denver, Colorado. Alexandria, VA: Geological Society of America 31(7): A-207.

DeGolyer and MacNaughton. 2001. 20th Century Petroleum Statistics, CD-ROM. Dallas, TX: DeGolyer and MacNaughton Worldwide Petroleum Consulting.

Dellagiarino, G., P. Fulton, K. Meekins, and D. Zinzer. 2000. Geological and Geophysical Data Acquisition: Outer Continental Shelf through 1999. Herndon, VA: U.S. Department of Interior, Minerals Management Service, Resource Evaluation Division, 31 pp.

Department of the Interior. 1999. Collection, Storage, Preservation and Scientific Study of Fossils from Federal and Indian Lands, background paper, 19 pp. Accessed December 12, 2001 at: http://www.fs.fed.us/geology/fedfos.pdf.

REFERENCES

DLF [Digital Library Federation]. 2002. Welcome to the Digital Library Federation. Accessed March 11, 2002 at: http://www.diglib.org/dlfhomepage.htm.

DOE [Department of Energy]. 1999. Oil and Gas R&D Programs. Accessed April 5, 2002 at: http://www.fe.doe.gov/oil_gas/progplan/99/istrat.pdf.

DOE [Department of Energy]. 2002. DOE Project Turns Abandoned Oil Lease into Million-Barrel Producer. Accessed March 20, 2002 at: http://www.fe.doe.gov/techline/tl_pru.shtml.

DOE [Department of Energy]. nd. Energy Glossary and Fact Sheets. Accessed March 27, 2002 at: http://www.eren.doe.gov/consumerinfo/glossary.html.

DOSECC [Drilling, Observation and Sampling of the Earth's Continental Crust, Inc]. 1998. Homepage. Accessed March 5, 2002 at: http://www.dosecc.org.

Eaton, G. P. 1996. What's ahead for the USGS? Geotimes 41(3):24-26.

Eguchi, R. T. 1998. Direct economic losses from the Northridge earthquake: A 3-year post-event perspective. Earthquake Spectra 14(2):245-264.

Faul-Zeitler, R. 1998. The Quandary of Federally Related Collections. Association of Systematics Collections, Washington, DC. Accessed on April 5, 2002 at: http://www.ascoll.org/Government/Government/quandry.htm.

FGDC [Federal Geographic Data Committee]. 2002. Homepage. Accessed March 11, 2002 at: http://www.fgdc.gov.

Geoinformatics Network. 2001. Geoinformatics: A Defining Opportunity for Earth Science Research Overview. Accessed March 12, 2002 at: http://www.geoinformaticsnetwork.org/overview.html.

Graves, R. W., and P. G. Sommerville. 1995. Characterizing long-period (1-10 sec.) ground motions for base isolated structures located in sedimentary basins. Pp. 59-65 in Seismic, Shock, and Vibration Isolation, vol. 319. Washington, DC: American Society of Mechanical Engineers Pressure Vessels and Piping.

Hoblitt R.P., J. S. Walder, C. L. Driedger, K. M. Scott, P. T. Pringle, and J. W. Vallance. 1998. Volcano Hazards from Mount Rainier, Washington, revised. USGS Open-File Report 98-428. Accessed January 23, 2002 at: http://vulcan.wr.usgs.gov/Volcanoes/Rainier/Hazards/OFR98-428/framework.html.

Hughes, N.C., F. J. Collier, J. Kluessendorf, J. H. Lipps, W. L. Taylor, and R. D. White. 2000. Institutional and individual orphaned collections. Pp. 25-36 in R. D. White and W. D. Allmon, eds. Guidelines for the Management and Curation of Invertebrate Fossil Collections: Including a Data Model and Standards for Computerization. Special Publications vol. 10. Ithaca, NY: Paleontological Society.

IHS Energy Group. 2002. Homepage. Accessed March 5, 2002 at: http://www.ihsenergy.com.

IMLS [Institute of Museum and Library Sciences]. 2002. Overview of the Museum and Library Services Act of 1996. Accessed March 11, 2002 at: http://www.imls.gov/whatsnew/leg/leg_mlsa.pdf.

IRIS [Incorporated Research Institutions for Seismology]. 2002. Homepage. Accessed March 5, 2002 at: http://www.iris.edu.

Iowa Department of Natural Resource. 2002. GEOSAM. Accessed February 27, 2002 at: http://gsbdata.igsb.uiowa.edu/geosam/.

Jackson, J. A., ed. 1997. Glossary of Geology, 4th ed. Alexandria, VA: American Geological Institute, 769 pp.

Johnson, K. R. 2001. The certification in paleontology program at the Denver Museum of Nature and Science. Abstracts with Programs from Annual Meeting in Boston, Massachusetts 3(6)34. Alexandria, VA: Geological Society of America.

Jones, C. I., and J. C. Williams. 1998. Measuring the social return to R&D. Quarterly Journal of Economics XX:1119-1135.

Jordan, J. 2000. Governors' natural gas summit: responding to the looming energy crisis, September 20. Washington, DC: Independent Petroleum Association of America, 4 pp.

Kansas Geological Survey. 2001. Kansas Wireline Log Header Database. Accessed March 11, 2002 at: http://magellan.kgs.ukans.edu/Elog/index.html.

Kentucky Geological Survey. 2001. Well Sample and Core Library. Accessed March 5, 2002 at: http://www.uky.edu/KGS/pubs/wellsamplelibrary.html.

Lane, N. G. 2000. Geology at Indiana University 1840-2000. Bloomington: Indiana University, 231 p.

Library of Congress. 1995. ANSI/NISO Z39.50-1995 [American National Standard Institute/National Information Standards Organization] Information Retrieval (Z39.50): Application Service Definition and Protocol Specification. Accessed March 12, 2002 at: http://lcweb.loc.gov/z3950/agency/markup/markup.html.

Library of Congress. 2002. Home Page of American Memory: Historical Collections for the National Digital Library. Accessed March 12, 2002 at: http://memory.loc.gov/.

Magistrale, H., S. Day, R. W. Clayton, and R. Graves. 2000. The SCEC Southern California reference three-dimensional seismic velocity model, version 2. Bulletin of the Seismological Society of America 90:565-567.

Marston, W. 1997. Jurassic mart: skull duggery and libertarian politics are driving the battles between paleontologists and commercial fossil traders. The Sciences 37(4):12-14.

Masters, J. A., ed. 1984. Elmworth—Case Study of a Deep Basin Gas Field. Memoir 38. Tulsa, OK: American Association of Petroleum Geologists, 316 pp.

McFarling, U. L. 2001. Ancient bone sales thrive in capitalist era. Fossils: auctioning of dinosaurs and other natural history relics angers scientists. Woolly mammoth tusk fetches $32,000 Sunday. Los Angeles Times, January 22. p. A1.

Montgomery, S. L. 1999. Core values: the growing need for repositories. Oil and Gas Journal Nov. 15:84-87.

Morell, V. 1998. A dinosaur for the mantel. Natural History 107(8):58-65.

Morris, P. J. 2000. A data model for invertebrate paleontological collections information. Pp. 105-108 in R. D. White and W. D. Allmon, eds. Guidelines for the Management and Curation of Invertebrate Fossil Collections: Including a Data Model and Standards for Computerization, special publications vol. 10. Ithaca, NY: Paleontological Society.

Morton-Thompson, D., and A. M. Woods, eds. 1993. Development Geology Reference Manual, Methods in Exploration Series, no. 10. Tulsa, OK: American Association of Petroleum Geologists, 550 pp.

NARA [National Archives and Records Administration]. 2002a. Appendix C: Appraisal Guidelines for Permanent Records. Accessed March 21, 2002 at: http://www.nara.gov/records/pubs/dfr/dfrappc.html.

NARA [National Archives and Records Administration]. 2002b. IV Record Values and Schedule Instructions. Accessed March 21, 2002 at: http://www.nara.gov/records/pubs/dfr/index.html.

National Energy Policy Development Group. 2001. National Energy Policy: Reliable, Affordable and Environmentally Sound Energy for America's Future. Washington, DC: Government Printing Office, 170 pp. Accessed February 4, 2002 at: http://www.whitehouse.gov/energy/National-Energy-Policy.pdf.

Natural History Museum, London. 2002. Catalog of Meteorites. Accessed March 11, 2002 at: http://www.nhm.ac.uk/mineralogy/grady/catalogue.htm.

NGDC [National Geophysical Data Center]. 2001. More about NGDC. Accessed March 11, 2002 at: http://www.ngdc.noaa.gov/ngdcinfo/aboutngdc.html.

NICL [National Ice Core Laboratory]. 2001. Master Inventory List. Accessed March 11, 2002 at: http://nicl.usgs.gov/master.htm.

NICL-SMO [National Ice Core Laboratory-Science Management Office]. 2000. Ice Core Distribution Policy. Accessed March 5, 2002 at: http://www.nicl-smo.sr.unh.edu.

NIST [National Institute of Standards and Technology]. 2001. The NIST Stone Test Wall. Accessed March 7, 2002 at: http://stonewall.nist.gov/CONTENT/Documen.htm.

NOAA [National Oceanic and Atmospheric Administration]. 2000. World Data Center Paleoclimatology Data Citations. Accessed March 12, 2002 at: http://www.ngdc.noaa.gov/paleo/citation.html.

NRC [National Research Council]. 1993. Toward a Coordinated Spatial

Data Infrastructure for the Nation. Washington, DC: National Academy Press, 171 pp.

NRC [National Research Council]. 1995a. Preserving Scientific Data on Our Physical Universe: A New Strategy for Archiving the Nation's Scientific Information Resources. Washington, DC: National Academy Press, 67 pp.

NRC [National Research Council]. 1995b. Study on the Long-term Retention of Selected Scientific and Technical Records of the Federal Government: Working Papers. Washington, DC: National Academy Press, 127 pp.

NRC [National Research Council]. 1996a. Maintaining Oil Production from Marginal Fields: A Review of the Department of Energy's Reservoir Class Program. Washington, DC: National Academy Press, 82 pp.

NRC [National Research Council] 1996b. Mineral Resources and Society: A Review of the U.S. Geological Survey's Mineral Resource Surveys Program Plan. Washington, DC: National Academy Press, 87 pp.

NRC [National Research Council]. 1998. NASA's Distributed Active Archive Centers. Washington, DC: National Academy Press, 232 pp.

NRC [National Research Council]. 1999a. Meeting U.S. Energy Resource Needs: The Energy Resources Program of the U.S. Geological Survey. Washington, DC: National Academy Press, 68 pp.

NRC [National Research Council]. 1999b. Hardrock Mining on Federal Lands Washington, DC: National Academy Press, 247 pp.

NRC [National Research Council]. 2000. 50 Years of Ocean Discovery: NSF 1950-2000. Washington, DC: National Academy Press, 269 pp.

NRC [National Research Council]. 2001. Future Roles and Opportunities for the U.S. Geological Survey. Washington, DC: National Academy Press, 179 pp.

NRC [National Research Council]. 2002. Coal Waste Impoundments: Risks, Responses, and Alternatives. Washington, DC: National Academy Press, 230 pp.

North Dakota State Geological Survey. 2001. North Dakota State Fossil Collection Catalog. Accessed February 27, 2002 at: http://www.state.nd.us/ndfossils/collections/Catalog.html.

NSF/EAR [National Science Foundation/ Division of Earth Sciences]. 2002. Statement of Guidelines, Division of Earth Sciences (EAR), National Science Foundation, for the implementation of the Foundation's Data Sharing Policy, April. Accessed on April 5, 2002 at: http://www.geo.nsf.gov/ear/EAR_data_policy_204.doc.

NSF/ONR [National Science Foundation / Office of Naval Research]. 2001. Data Management for Marine Geology and Geophysics. Accessed January 23, 2002 at: http://humm.whoi.edu/DBMWorkshop/data_mgt_report.hi.pdf.

ODP [Ocean Drilling Program]. 2002. Janus Database. Accessed March 11, 2002 at: http://www-odp.tamu.edu/database/janusmodel.htm.

Office of Surface Mining. 1997. Mine Map Repositories: A Source of Mine Data. Washington, DC: Department of the Interior, 8 pp.

Pojeta, J. Jr. 2000. The more eyes the better: fossils and amateurs. Geotimes 45(10):5.

Razand, J., and P. E. Stutzman. 2001. Building Stone of the United States: The NIST Test Wall. Accessed April 5, 2002 at: http://stonewall.nist.gov/Default.htm.

Rivero, C., J. H. Shaw, and K. Mueller. 2000. Oceanside and Thirtymile Bank blind thrusts: implications for earthquake hazards in coastal southern California. Geology 28:891-894.

Schlumberger Ltd. 2002. Oilfield Glossary. Accessed April 5, 2002 at: http://www.glossary.oilfield.slb.com/Default.cfm.

Schweitzer, M. H., M. Marshal, K. Carron, D. S. Bohle, S. C. Busse, E. V. Arnold, D. Barnard, J. R. Horner, and J. R. Starkey. 1997. Heme compounds in dinosaur trabecular bone. Proceedings National Academy Sciences 94:6291-6296.

Secretary of the Interior. 2000. Fossils on Federal and Indian Lands. Washington, DC: Department of the Interior, 49 pp.

Serra, O. 1984. Fundamentals of Well-log Interpretation—the Acquisition of Logging Data. New York: Elsevier, 423 pp.

Shaw, J. H., and P. Shearer. 1999. An eluvise blind-thrust fault beneath metropolitan Los Angeles. Science 283:1516-1518.

Sheriff, R. E. 1994. Encyclopedia Dictionary of Exploration Geophysics. Third edition. Tulsa, OK: Society of Geophysicists, 323 pp.

Shovers, B., M. Fiege, D. Martin, and F. Quivik. 1991. Butte and Anaconda Revisited, Special Publication 99. Butte, MT: Montana Bureau of Mines and Geology, 63 pp.

Simpson, S. 2000. Bidding on bones. Scientific American, March. Accessed April 6, 2002 at: http://www.sciam.com/techbiz/0300techbus1.html.

Sledge, J. 1998. Protecting a "snapshot of the past." USDA News 57:4. Accessed December 2001 at: http://www.usda.gov/news/pubs/newslett/old/vol57no4/article5.htm.

Smithsonian Institution. 2001. The Creation of the Science Commission. Accessed March 4, 2002 at: http://www.si.edu/sciencecommission.

Sneider, R. M., and J. S. Sneider. 2001. New oil in old places. Pp. 63-84 in M. W. Downy, J. C. Threet, and W. A. Morgan, eds., Petroleum Provinces of the Twenty-first Century, Memoir 74. Tulsa, OK: American Association of Petroleum Geologists.

Stocur, J. G. 1986. The potential of enhanced oil recovery. International Journal of Energy Research 10:357-370.

Taylor, B. W. 1992. Cataloging: the first step to data management. Journal of the Society of Petroleum Engineering 2440:193-197.

Toner, M. 2001. Brazen fossil hunters are cleaning out U.S. dinosaur heritage. Atlanta Journal-Constitution, August 24, p. D-1.

Tsutsumi, H., R. S. Yeats, and G. J. Huftile. 2001. Late Cenozoic tectonics of the northern Los Angeles fault system, California. Geological Society of America Bulletin 113:454-468.

UCMP [University of California, Museum of Paleontology, Berkeley]. 2002a. Orphaned Collections. Accessed March 5, 2002 at: http://www.ucmp.berkeley.edu/ICAL/.

UCMP [University of California, Museum of Paleontology, Berkeley]. 2002b. Homepage. Accessed March 12, 2002 at: http://www.ucmp.Berkeley.edu.

University of Chicago Press. 2002. The Chicago Style Manual FAQ (and not so FAQ). Accessed February 27, 2002 at: http://www.press.uchicago.edu/Misc/Chicago/cmosfaq.html.

USGCRP [U.S. Global Change Research Program]. 1991. Policy Statements on Data Management for Global Change Research. Accessed March 5, 2002 at: http://www.gcrio.org/USGCRP/DataPolicy.html.

USGS [U.S. Geological Survey]. 1996. USGS Response to an Urban Earthquake: Northridge 1994. Open-File Report 96-263. Denver, CO: USGS, 77 pp.

USGS [U.S. Geological Survey]. 2002. Suggestions to Authors of Reports to the United States Geological Survey. Accessed February 27, 2002 at: http://www.nwrc.usgs.gov/lib/lib_sta.htm.

USGS [U.S. Geological Survey]. nd. USGS Central Region Geologic Information, Denver Library. Accessed on March 7, 2002 at: http://geology.cr.usgs.gov/crg/library.htm.

USGSA [U.S. General Services Administration]. 2002. E-gov. Accessed March 11, 2002 at: http://egov.gov.

Veritas. 2002. Homepage. Accessed March 11, 2002 at: http://www.veritasdgc.com.

VMNH [Virginia Museum of Natural History]. 2001. Invertebrate Paleontological Collection. Accessed March 11, 2002 at: http://www.vmnh.org/inverfos.htm.

WAIS [West Antarctic Ice Sheet]. 2000. Flow History of Siple Dome. Accessed March 5, 2002 at: http://www.geophys.washington.edu/Surface/Glaciology/PROJECTS/SIPLE/siple.html.

White, R. D., and W. D. Allmon. 2000. Guidelines for the Management and Curation of Invertebrate Fossil Collections. Ithaca, NY: Paleontological Society, 260 pp.

WOGCC [Wyoming Oil and Gas Conservation Commission]. 2001a. Rules. Chapter 3. Section 21. Filing of Well Logs. Accessed March 11, 2002 at: http://wogcc.state.wy.us/db/rules/3-21.html.

WOGCC [Wyoming Oil and Gas Conservation Commission]. 2001b. WOGCC Homepage. Accessed March 11, 2002 at: http://wogcc.state.wy.us.

Yeats, R. S., and G. J. Huftile. 1995. The Oak Ridge fault system and the 1994 Northridge earthquake. Nature 373:418-420.

Appendixes

A

Biographical Sketches of Committee Members

Christopher G. Maples, *Chair,* is the Chair of the Department of Geological Sciences at Indiana University. A paleontologist, his current research generally involves field- and literature-based studies of invertebrates or invertebrate traces. He uses these data to address questions that link paleontology and geology. Current research also includes several projects on Late Devonian through Triassic echinoderm extinction, extinction rebound, and biogeography from various parts of the world. He was the 1994 recipient of the Schuchert Award of the Paleontological Society. Dr. Maples has served in a number of academic, curational, research administrative, and geological survey administrative positions. Dr. Maples has served as program director for the Geology and Paleontology Program at the National Science Foundation, as chief of the Geologic Investigations Section of the Kansas Geological Survey (KGS) and assistant chief of the Petroleum Research Section of the KGS. Dr. Maples' professional and advisory activities include serving as president of the board of trustees for the Paleontological Research Institution; chair of American Geophysical Union Geoscience Heads and Chairs; councilor for the Paleontological Society; and associate editor for various professional journals in geology and paleontology.

Warren D. Allmon is the director of the Paleontological Research Institution and an adjunct associate professor in the Department of Earth and Atmospheric Sciences at Cornell University. His major research interest is the ecology of the origin and maintenance of biological diversity and the application of the geological record to the study of these problems. Most recently his work includes research on the paleoceanography and paleoclimate of the western Atlantic Ocean during the last 10 million to 20 million years, and the possible influence of paleoceanographic conditions on the evolution of mollusks in this region. Dr. Allmon has edited several texts and published numerous scientific and popular articles. He is a fellow of the Geological Society of America.

Kevin Thomas Biddle received a Ph.D. in geology from Rice University. He is vice president-South America at ExxonMobil Exploration Company. He has been with Exxon—now ExxonMobil—since 1978 and has held numerous positions of increasing responsibility, including: ventures manager, West Africa and Far East; operations manager, West Africa; divisions manager, supervisor, and geological advisor. In 1973 and 1974, Dr. Biddle worked for the U.S. Geological Survey in Menlo Park on projects in Alaska and Southern California. He is involved in several professional activities, including serving as the elected editor of the *Bulletin of the American Association of Petroleum Geologists* for 4 years, and is a fellow of the Geological Society of America, and a member of the American Geophysical Union, the American Association of Petroleum Geologists, and the Houston Geological Society. Dr. Biddle has served as a member of two NRC committees, the Panel on the Geodynamics of Sedimentary Basins and the U.S. Geodynamics Committee.

Donald D. Clarke is division engineer and chief geologist with the Department of Oil Properties, City of Long Beach, California, and teaches geology at Compton Community College. He received his Bachelor's degree in geology from California State University–Northridge, with additional graduate study at California State University–Northridge, –Los Angeles, and –Long Beach. Mr. Clarke began his career in 1974 as an energy and mineral resources engineer with the California State Lands Commission. He worked extensively on the giant Wilmington oil field and the California offshore. Since 1981 he has been with the City of Long Beach Department of Oil Properties where he is Division Engineer for Geology, Environment and Safety. A member of the Los Angeles Basin Geologic Society since 1974, he has served as president from 1996 to 2001. Over the years he has focused on community outreach and education. Mr. Clarke has also served as chairman of the Long Beach Unit Equity Geology and Sand Volume Subcommittees. A member of AAPG since 1986, he is currently on the Advisory Council representing the Pacific Section, and on the Advisory Board for the Division of Environmental Geosciences. He is also on the AAPG Standing Committees on Public Information, and on Reservoir Development and he received the AAPG Distinguished Service Award in 2002. Mr. Clarke has published or presented more than 50 technical papers on topics that include computer mapping, sequence stratigraphy, horizontal drilling, structural geology, and reservoir evaluation, and he has been recognized by the Institute for the Advancement of Engineering as a fellow.

Beth Driver is the scientific advisor for data bases at the National Imagery and Mapping Agency. Her responsibilities have focused on defining an information architecture to support generation, reuse, and dissemination of a dynamic set of imagery and mapping products and on exploring new methods for acquiring and producing geospatial data. Dr. Driver has led various projects exploring alternative business practices, such as the current effort to enhance intelligence and geospatial support for precision strike operations and to enhance the geospatial accuracy of imagery intelligence reporting. Other projects have explored application of new technology to the production and dissemination of NIMA data. Dr. Driver has served on national standards boards for relational data base management systems and for geospatial data. Prior to joining the Defense Mapping Agency, Dr. Driver managed system engineering and system development efforts for intelligence customers of the defense contractor community. She also served as a member of the NRC's Panel to Review the Oak Ridge Active Archive Center (DAAC).

Thomas R. Janecek is curator of the Antarctic Marine Geology Research Facility at Florida State University. His research interests include paleoceanography, paleoclimatology, and deep-sea sedimentology. His previous positions include terms at the Lamont-Doherty Earth Observatory of Columbia University as a research scientist and the Ocean Drilling Program as a staff scientist. He has sailed on 14 deep-sea drilling expeditions and has participated in an international coring project in Antarctica. He is currently involved in U.S. and international efforts to establish long-term, scientific deep-sea drilling and coring programs in the southern oceans surrounding Antarctica and in the Arctic Ocean.

Linda R. Musser is head of the Earth and Mineral Sciences Library at the Pennsylvania State University. Ms. Musser received her B.S. in civil engineering and worked for the U.S. Army Corps of Engineers and in industry. She has an M.S. in library and information science from the University of Illinois; her research relates to the challenges and methods for preserving and providing access to scientific and technical information. She is a member of the American Library Association, Special Libraries Association, American Society for Engineering Education, Tau Beta Pi, and the Geoscience Information Society. She is past president of the Engineering Libraries Division of the American Society for Engineering Education and served as co-chair of the Geoscience Information Society Preservation Committee.

Robert W. Schafer is a mineral exploration and business development consultant. From 1996 to 2002 he was vice president for exploration at Kinross Gold Corporation. Over the 24-year period before working with Kinross, Mr. Schafer worked as regional exploration manager for BHP Minerals, exploration manager for Addwest Gold Corporation and Billiton Metals, and as an exploration geologist for U.S. Borax-RTZ. He is active in the Society of Economic Geologists, where he is on the Executive Council and a trustee on its Foundation Board. He is a member of the Society for Mining, Metallurgy and Exploration (SME), where he sits on the Board of Directors; Northwest Mining Association, where he held a position as trustee; and the Mining and Metallurgical Society of America, where he is currently vice president. Mr. Schafer is also active in the Prospectors and Developers Association as a board member, and Canadian Institute of Mining where he has been on the planning committees for the 1999, 2000, and 2001 meetings. He is a past president of the Geological Society of Nevada and has authored and edited a number of publications concerning mineral exploration and the business of exploration. In 2002, the American Institute of Mining, Metallurgical and Petroleum Engineers selected Mr. Schafer to receive the William Saunders Gold Medal.

Robert M. Sneider (NAE) is president of Robert M. Sneider Exploration, Inc. Dr. Sneider received his Ph.D. from the University of Wisconsin. His primary interests include petroleum exploration, property acquisition, and integrated geoscience-petroleum engineering studies. For the past 35 years, he has studied the geological, petrophysical, and engineering properties of petroleum reservoir and seal rocks. Prior to 1981, Dr. Sneider was a partner with Sneider and Meckel Associates, Inc., a geological, geophysical, and petroleum engineering consulting and exploration company. He also spent 17 years with Shell Oil and Shell Development in a number of different areas of geology and petrophysics. He is a member of National Academy of Engineering Section 11: Petrology, Mining and Geologic Engineering. Dr. Sneider is an honorary member of the American Association of Petroleum Geologists and a recipient of their Sidney Powers medal.

John C. Steinmetz is director of the Indiana Geological Survey and State Geologist of Indiana. Before coming to Indiana in 1998, Dr. Steinmetz held a similar position at the Montana Bureau of Mines and Geology for 4 years. His previous positions include terms on the faculty of the Department of Marine Science, University of South Florida; adjunct professor of Geology, University of Montana; and senior research geologist for Marathon Oil Company in Littleton, Colorado. He is currently the treasurer of the Association of American State Geologists and secretary to the Executive Committee of the American Geological Institute (AGI). He is a member of the GeoRef Advisory Board for the AGI, the Board of Trustees of the Paleontological Research Institution, and the Advisory Board of Micropaleontology Press. His research interests include biostratigraphy, micropaleontology, and the geology of Indiana. Dr. Steinmetz received his Bachelor's and Master's degrees in geology from the University of Illinois; he earned his Ph.D. in marine geology and geophysics from the Rosenstiel School of Marine and Atmospheric Science, University of Miami.

Sally Zinke is a geophysical consultant in Denver, Colorado, involved in a number of U.S. and international projects. Ms. Zinke joined Mobil in 1973 and held a number of technical and managerial positions in exploration and production. Ms. Zinke also handled geophysical applications for the Bureau of Economic Geology, University of Texas, for several reservoir characterization and technology transfer integration projects. She was a geophysical coordinator for development and implementation of relational database software for a consortium of 15 major international upstream petroleum companies. Her main interests are technology integration at the reservoir level, and high resolution seismic and reservoir characterization. Ms. Zinke has served as chair for a number of Society of Exploration Geophysicists committees and formerly held the office of president. She is an active member of American Association Petroleum Geologists, European Association of Geoscientists and Engineers, Denver Geophysical Society, and the Rocky Mountain Association of Geologists.

NRC Staff

Paul Cutler, *study director,* is a program officer at the NRC Board on Earth Sciences and Resources. He received a Bachelor's degree from Manchester University, England, a Master's degree from the University of Toronto, and a Ph.D. from the University of Minnesota. Prior to joining the NRC, Dr. Cutler was an assistant scientist and lecturer in the Department of Geology and Geophysics at the University of Wisconsin–Madison. His research is in surficial processes, specifically glaciology, hydrology, and Quaternary science. In addition to numerical modeling and GIS-based research, he has conducted field studies in Alaska, Antarctica, Arctic Sweden, the Swiss Alps, Pakistan's Karakoram Mountains, the midwestern United States, and the Canadian Rockies. He is a member of the Geological Society of America, the American Geophysical Union, and the Geological Society of Washington, and a fellow of the Royal Geographical Society.

Monica Lipscomb is a research assistant for the NRC Board on Earth Sciences and Resources. She is completing a Master's of Urban and Regional Planning at Virginia Polytechnic Institute. Previously, she served as a Peace Corps volunteer in the Côte d'Ivoire and has worked as a biologist at the National Cancer Institute. She holds a B.S. in environmental and forest biology from the State University of New York—Syracuse.

B

Presentations to the Committee

ORAL PRESENTATIONS AND STATEMENTS

Wayne Ahr, Texas A&M, College Station
Edith Allison, Department of Energy, Washington, D.C., and American Association of Petroleum Geologists, Tulsa, Oklahoma
M. Lee Allison, Kansas Geological Survey, Lawrence
David Archer, Petrochemical Open Software Corporation, Houston, Texas
Richard Benson, National Museum of Natural History, Washington, D.C.
Carolyn Bertrand, Ocean Energy, Dallas, Texas
Glenn Breed, UpstreamInfo, Houston, Texas
Ronald Broadhead, New Mexico Bureau of Geology and Mineral Resources, Socorro, New Mexico
Lawrence Bruno, Core Labs, Houston, Texas
George Bush, Bureau of Economic Geology, University of Texas at Austin
Elizabeth Campbell, Energy Information Administration, Washington, D.C.
Stewart Chuber, South Texas Geological Survey, San Antonio, Texas
Bill Cobbin, USGS *emeritus,* Denver, Colorado
Eric Cravens, National Ice Core Laboratory, Denver, Colorado
David Davies, Geosystems, Houston, Texas
Jimmy Denton, BP Amoco, Tulsa, Oklahoma
David DeWitt, South West Florida Water Management District, Tampa, Florida
William Dirks, Samson Resources, Houston, Texas
David Divins, National Geophysical Data Center, Boulder, Colorado
Charles Downs, National Archives and Records Administration, College Park, Maryland
James Edwards, Bureau of Land Management, Lakewood, Colorado
William Fisher, Bureau of Economic Geology, University of Texas at Austin, *emeritus*
Joan Fitzpatrick, U.S. Geological Survey, Denver, Colorado
Charles Groat, U.S. Geological Survey, Reston, Virginia
Linda Gundersen, U.S. Geological Survey, Reston, Virginia
Leslie Hale, National Museum of Natural History, Washington, D.C.
Geoffrey Hargreaves, National Ice Core Laboratory, Denver, Colorado
Christopher Keane, American Geological Institute, Alexandria, Virginia
Joseph King, National Aeronautics and Space Administration, Greenbelt, Maryland
Michael Klosterman, U.S. Army Corps of Engineers, Washington, D.C.
Michael Kurtz, National Archives and Records Administration, College Park, Maryland
John Ladd, Kerr McGee Rocky Mountain Corporation, Denver, Colorado
Susan Landon, Thomasson Partner Associates, Denver, Colorado
H. Richard Lane, National Science Foundation, Arlington, Virginia
Chris Lewis, Yucca Mountain Site Characterization Project, Mercury, Nevada
Bill Linenberger, USGS Core Research Center, Denver, Colorado
Susan Longacre, ChevronTexaco, Houston, Texas
Mark Longman, Carbonate Rocks and Reservoirs, Lakewood, Colorado
Gary Lore, Mineral Management Service, New Orleans, Louisiana
Howard Lowell, National Archives and Records Administration, College Park, Maryland

Charles Mankin, Oklahoma Geological Survey, Norman, Oklahoma
Robert Martin, Institute of Museum and Library Services, Washington, D.C.
Randi Martinsen, Institute for Energy Research, Laramie, Wyoming
Richard Marvel, Wyoming Oil and Gas Conservation Commission, Casper, Wyoming
Timothy McCoy, National Museum of Natural History, Washington, D.C.
Kevin McKinney, U.S. Geological Survey, Denver, Colorado
Laura Mercer, Denver Earth Resources Library, Denver, Colorado
Robert Merrill, Samson Resources, Tulsa, Oklahoma
Tom Michalski, USGS Core Research Center, Denver, Colorado
Michael Miller, National Archives and Records Administration, College Park, Maryland
Marcus Milling, American Geological Institute, Alexandria, Virginia
Allan Montgomery, USGS, Reston, Virginia
Carla Moore, National Geophysical Data Center, Boulder, Colorado
David Morehouse, Energy Information Administration, Washington, D.C.
Steve Natali, Barrett Resources, Denver, Colorado
Dennis Nielson, DOSECC, Inc., Salt Lake City, Utah
Daniel Ortuño, Bureau of Economic Geology, University of Texas at Austin
Michael Padgett, EEX Corporation, Houston, Texas
Julie Palais, National Science Foundation, Arlington, Virginia
James Parr, Anadarko Petroleum, Houston, Texas
Don Paul, ChevronTexaco, San Francisco, California
Richard Pawlowicz, Bechtel National Inc., San Diego, California
Frank Rack, Joint Oceanographic Institutions, Washington, D.C.
Douglas Ratcliff, Bureau of Economic Geology, University of Texas at Austin
Katie Ryan, Library of Congress, Washington, D.C.
Ron Samuels, IHS Resources, Houston, Texas
Robert Shafer, C&M Storage, Schulenberg, Texas
George Sharman, National Geophysical Data Center, Boulder, Colorado
Sally Shelton, National Museum of Natural History, Washington, D.C.
Sorena Sorensen, National Museum of Natural History, Washington, D.C.
Duane Spencer, Bureau of Land Management, Lakewood, Colorado
Susan Steele Weir, Denver Water Department, Denver, Colorado
Emily Stoudt, ChevronTexaco, Midland, Texas
James Stouffer, Alaska Department of Natural Resources, Anchorage
Kenneth Telchik, Internal Revenue Service, Dallas, Texas
Philippe Theys, Schlumberger Inc., Sugarland, Texas
Kenneth Thibodeau, National Archives and Records Administration, College Park, Maryland
Jann Thompson, National Museum of Natural History, Washington, D.C.
Scott Tinker, Bureau of Economic Geology, University of Texas at Austin
William Trapmann, Energy Information Administration, Washington, D.C.
Kay Waller, Denver Earth Resources Library, Denver, Colorado
Robert Weimer, Colorado School of Mines, Golden, Colorado
Anna Weitzman, National Museum of Natural History, Washington, D.C.
Guenter Wellmann, Alberta Core Repository, Calgary, Alberta, Canada
Bill Whitus, Core Research Center, Denver, Colorado
Yvonne Wilson, National Archives and Records Administration, College Park, Maryland
Scott Wing, National Museum of Natural History, Washington, D.C.
John Wiltshire, University of Hawaii, Honolulu
Mel Wright, City of Long Beach (retired), California
Herman Zimmerman, National Science Foundation, Alexandria, Virginia

WRITTEN STATEMENTS TO THE COMMITTEE

William Akersten, Idaho State University, Pocatello
Elinor Alexander, Mineral and Energy Resource Department, Adelaide, South Australia
Edith Allison, Department of Energy, Washington, D.C., and American Association of Petroleum Geologists, Tulsa, Oklahoma
Greg B. Arehart, University of Nevada, Reno
Heather Astwood, Nuclear Regulatory Commission, Washington, D.C.
Ron Baker, Petroleum Extension Service, University of Texas at Austin
Paul J. Bartos, Colorado School of Mines, Golden
Larry Baume, National Archives and Records Administration, College Park, Maryland
Richard Benson, National Museum of Natural History, Washington, D.C.
Don Birak, Anglo Gold Corporation, Cripple Creek and Victor, Colorado
Robert Blodgett, Oregon State University, Corvallis
Randy Bolles, Wyoming Department of Revenue, Cheyenne
Arthur Boucot, Oregon State University, Corvallis

David W. Brown, State Lands Commission, Sacramento, California
Paul L. Brown, Bureau of Land Management, Denver, Colorado
Rex Buchanan, Kansas Geological Survey, Lawrence
George Bush, Bureau of Economic Geology, University of Texas at Austin
Timothy Carr, Kansas Geological Survey, Lawrence
Steve Carey, University of Rhode Island, Kingston
Robert Chase, Millstream Mines, Ltd., Toronto, Ontario, Canada
Stewart Chuber, Fayette Exploration, Schulenberg, Texas
Brad Cook, Indiana University Archives, Bloomington
Jimmy Denton, BP Amoco, Tulsa, Oklahoma
Elaine Dobinson, NASA Jet Propulsion Laboratory, Pasadena, California
Doug Erwin, Smithsonian National Museum of Natural History, Washington, D.C.
Dorothy Ettensohn, Natural History Museum of Los Angeles, California
John Firth, Ocean Drilling Program, College Station, Texas
William Fisher, Bureau of Economic Geology, University of Texas at Austin, *emeritus*
Rita Frasure, Northern Rockies Geologic Data Center, Billings, Montana
Charles Gaines, USDA ARS Soft Wheat Quality Laboratory, Wooster, Ohio
Ronald Gashinski, Ontario Geologic Survey, Canada
Susan Granados, Florida Atlantic University, Boca Raton
Elwyn Griffiths, ExxonMobil, Houston, Texas
Jake Hancock, Imperial College of Science, Technology and Medicine (*emeritus*), London, England
Geoffrey Hargreaves, National Ice Core Laboratory, Denver, Colorado
Ernest W. Harrison, Millstream Mines, Ltd., North Bay, Ontario, Canada
Donald Hartman, Devon Energy, Houston, Texas
Robert Hunt, University of Nebraska, Lincoln
Shepley Jackson, ExxonMobil Upstream Research Company, Houston, Texas
Kathleen Johnson, U.S. Geological Survey, Reston, Virginia
Kirk Johnson, Denver Museum of Nature and Science, Colorado
Philip Justus, Nuclear Regulatory Commission, Rockville, Maryland
Roger L. Kaesler, University of Kansas, Lawrence
Anthony Kampf, Natural History Museum of Los Angeles, California
Patricia Kelley, University of North Carolina, Wilmington
Harold Kemp, Wyoming Office of State Lands and Investments, Cheyenne
Jim Kennedy, Oxford Museum of Natural History, Oxford, England
Richard Ketelle, Bechtel Jacobs, Oakridge, Tennessee

Joseph King, NASA Goddard Space Flight Center, Greenbelt, Maryland
John Kingsolver, Chicago Jewish Archives, Illinois
Michael Klosterman, U.S. Army Corps of Engineers, Washington, D.C.
Kjell Reidar Knudsen, Norwegian Petroleum Directorate, Stavanger, Norway
Richard Lane, National Science Foundation, Arlington, Virginia
Joel Lardon, Bureau of Economic Geology, University of Texas at Austin
Chris Lewis, Department of Energy, Las Vegas, Nevada
Don Lewis, American Association of Petroleum Geologists, Tulsa, Oklahoma.
Donald Lindsay, Consultant, Bakersfield, California
Brian Logan, Minerals and Energy Resource Department, Adelaide, South Australia
Cindy Lohman, National Science Foundation, Arlington, Virginia
Mary Jo Lynch, American Library Association, Chicago, Illinois
Sandra Mark, Petroleum Technology Transfer Council, Denver, Colorado
Tim McCoy, Smithsonian National Museum of Natural History, Washington, D.C.
Robert Merrill, Sampson Resources, Houston, Texas
David L. Meyer, University of Cincinnati, Ohio
John Mitchell, Northern Rockies Geologic Data Center, Billings, Montana
David Morehouse, Energy Information Administration, Washington, D.C.
Keiko Motoyama, Kongo Shelves, Ltd., Tokyo, Japan
Claudia M. Newbury, U.S. Department of Energy, North Las Vegas, Nevada
David Nicklin, ARCO (*emeritus*), Laguna Niguel, California
Michael Padgett, EEX Corporation, Houston, Texas
Larry Page, National Science Foundation, Arlington, Virginia
Julie Palais, National Science Foundation, Arlington, Virginia
Allison R. Palmer, Institute for Cambrian Studies, Boulder, Colorado
Tim Palmer, University of Wales, Aberystwyth
Debbie Patskowski, Alaska Geological Materials Center, Eagle River
John Pojeta, Smithsonian, National Museum of Natural History, Washington, D.C.
Frank Rack, Joint Oceanographic Institutions, Washington, D.C.
Emma C. Rainforth, Columbia University, New York City, New York
Robert Raynolds, Denver Museum of Nature and Science, Colorado

APPENDIX B

John Reeder, Alaska Geological Materials Center, Eagle River
Greig Robertson, National Mine Map Repository, Pittsburgh, Pennsylvania
Peter Robinson, University of Colorado Museum of Natural History, Boulder
Reuben Ross, Colorado School of Mines, Littleton
John Shaw, Harvard University, Cambridge, Massachusetts
Andrew Sicree, Pennsylvania State University, University Park
Ross B. Simons, Smithsonian, National Museum of Natural History, Washington, D.C.
Jann Thompson, Smithsonian, National Museum of Natural History, Washington, D.C.
Natalie Uschner, Indiana University, Bloomington
Ireneusz Walaszczyk, University of Warsaw, Poland
Suzanne Weedman, USGS, Reston, Virginia
Richard B. Wheeler, ExxonMobil Upstream Research Company, Houston, Texas
Tim White, Yale University, New Haven, Connecticut
Sherilyn C. Williams-Stroud, ChevronTexaco, Bellaire, Texas
Ted Wilton, Kinross Gold, Salt Lake City, Utah
Wendy Wiswall, U.S. National Museum of Natural History, Washington, D.C.
Thomas L. Wright, Chevron (*emeritus*), San Anselmo, California
Herman B. Zimmerman, National Science Foundation, Arlington, Virginia
Andrew Zolnowski, Killam Associates, Millburn, New Jersey

RESPONDENTS TO POLL ABOUT DATA AND COLLECTIONS AT STATE GEOLOGICAL SURVEYS

Alabama
Donald F. Oltz and W. Edward Osborne, Geological Survey of Alabama

Alaska
Milton A. Wiltse, Alaska Division of Geological and Geophysical Surveys

Arizona
Larry D. Fellows and Steven L. Rauzi, Arizona Geological Survey

California
James F. Davis and David J. Beeby, California Division of Mines and Geology

Colorado
Vicki J. Cowart, Colorado Geological Survey

Connecticut
Ralph Lewis, Connecticut Geological and Natural History Survey

Delaware
Robert R. Jordan, Delaware Geological Survey

Florida
Walter Schmidt and Thomas M. Scott, Florida Geological Survey

Hawaii
Glenn R. Bauer, Department of Land and Natural Resources

Illinois
William W. Shilts, Donald E. McKay, and Marie-France Dufour, Illinois State Geological Survey

Indiana
John A. Rupp, Indiana Geological Survey

Iowa
Donald L. Koch and Raymond Anderson, Iowa Geological Survey Bureau

Kansas
M. Lee Allison, William Harrison, and Barbara McClain, Kansas Geological Survey

Kentucky
James C. Cobb, Kentucky Geological Survey

Louisiana
Chacko J. John, Louisiana Geological Survey

Maine
Robert G. Marvinney, Maine Geological Survey

Massachusetts
Richard N. Foster, Massachusetts Geological Survey

Michigan
Harold R. Fitch and Steven E. Wilson, Michigan Geological Survey Division

Minnesota
David L. Southwick and Dale R. Setterholm, Minnesota Geological Survey

Missouri
Mimi R. Garstang and Ardel Rueff, Missouri Division of Geology and Land Survey

Montana
Edmond G. Deal, Richard B. Berg, Robin B. McCulloch, and Thomas W. Patton, Montana Bureau of Mines and Geology

Nebraska
Mark S. Kuzila, Nebraska Conservation and Survey Division

Nevada
Jonathan G. Price, Nevada Bureau of Mines and Geology

New Hampshire
David R. Wunsch, Department of Environmental Services

New Mexico
Peter A. Scholle, and Ronald Broadhead, New Mexico Bureau of Geology and Mineral Resources

New York
Robert H. Fakundiny and Robert H. Fickies, New York State Geological Survey
William M. Kelly, Curator, Mineralogy Collection, New York State Museum

North Carolina
Charles H. Gardner and Kenneth B. Taylor, North Carolina Geological Survey

North Dakota
John P. Bluemle, Mark A. Gonzalez, and Julie A. LeFever, North Dakota Geological Survey

Ohio
Thomas M. Berg, Ohio Division of Geological Survey

Oklahoma
Charles J. Mankin, Oklahoma Geological Survey

Oregon
John Beaulieu and Dennis Olmstead, Oregon Department of Geology and Mineral Industries

South Dakota
Derric L. Iles, South Dakota Geological Survey

Tennessee
Ronald P. Zurawski and Marvin Berwind, Tennessee Division of Geology

Texas
Scott W. Tinker and Douglas C. Ratcliff, Bureau of Economic Geology, University of Texas at Austin

Utah
Rick Allis, Christine Wilkerson, and Carolyn Olsen, Utah Geological Survey

Virginia
Stanley Johnson and Palmer Sweet, Virginia Division of Mineral Resources

West Virginia
Larry D. Woodfork, West Virginia Geological and Economic Survey

Wisconsin
James M. Robertson and Thomas J. Evans, Wisconsin Geological and Natural History Survey

Wyoming
Lance Cook, Wyoming State Geological Survey

RESPONDENTS TO POLL ABOUT DATA AND COLLECTIONS AT OTHER INSTITUTIONS

Alberta Energy and Utilities Board Core Research Center
Guenter Wellmann, Alberta EUB, Calgary

Bureau of Economic Geology Core Research Center
Douglas Ratcliff, Bureau of Economic Geology, University of Texas at Austin

Bureau of Economic Geology GLF
Daniel Ortuño, Bureau of Economic Geology, University of Texas at Austin

California Well Sample Repository
Larry Knauer, California Well Sample Repository, Bakersfield

C&M Storage, Inc.
Robert Shafer, C&M Storage, Inc., Schulenberg

Denver Earth Resources Library
Kay Waller and Laura Mercer, Denver Earth Resources Library, Denver

Los Angeles Basin Subsurface Data Center
Daniel Francis, Los Angeles Basin Subsurface Data Center
Stanley Finney, Los Angeles Basin Subsurface Data Center

Los Angeles County Museum of Natural History
Anthony Kampf and Dorothy Ettensohn, Natural History Museum of Los Angeles County

APPENDIX B

National Archives and Records Administration
Howard Lowell, NARA, College Park

National Geophysical Data Center Marine Geology and Geophysics Program
Carla Moore, NGDC-MGG, Boulder
David Divens, NGDC-MGG, Boulder

National Ice Core Laboratory
Geoffrey Hargreaves, NICL, Lakewood

National Lacustrine Core Repository
Douglas Schnurrenberger and Linda Shane, University of Minnesota, Minneapolis

Ocean Drilling Program
John Firth, Ocean Drilling Program, College Station, Texas

Smithsonian Institution
Ross Simons, Smithsonian Institution, Washington, D.C.

U.S. Army Corps of Engineers
Michael Klosterman, USACE, Washington, D.C.

U.S. Geological Survey Paleontology Collection
Kevin McKinney, USGS, Denver

U.S. Geological Survey Core Research Center
Thomas Michalski, USGS, Denver

University of Rhode Island
Steven Carey, University of Rhode Island, Narragansett

C

Questionnaire

Please note that we aim to ask a uniform set of questions of all facilities. The use of the word "thing" is one attempt to make the list of questions universally applicable!

1. Size.
   a. How many [things] are in the collection? (Whatever unit of measurement you use: perhaps number, volume, area, etc).
   b. How do you evaluate your holdings?
   c. How fast is the collection growing?
   d. How much more space is available? How much total space existed in the first place?
   e. Have you lost any holdings due to disasters (i.e. flooding, fire, etc)?

2. Usage.
   a. What sorts of people use the [things] in a given period of time? What for?
   b. How many people use the [things] in a given period of time?
      For example, How many people used the catalog?
      How many people used specimens?
      How many people requested data?
   c. What is the long-term trend in usage of your holdings, and does this differ between individual data types?
   d. Can you provide any notable examples of people using your holdings for purposes other than those they were initially collected for?
   e. What might encourage greater use of your holdings?

3. Accessibility.
   a. Are all of the [things] accessible at any given point in time?
   b. How do patrons get access to your data?
   c. Could you provide sample catalog entries for your specimens and for your data.
   d. How do you prioritize your cataloging?

4. Data Management.
   a. What transcription programs do you have for your data at present?
      For example: Digitizing hardcopy records:
      Copying from one electronic medium to another:
      Re-formatting data:

5. Accession/de-accession.
   a. Who is giving you material?
   b. What are your accession/de-accession protocols?
   c. What data standards (of any kind) do you use?
   d. Do you de-accession? How much? How many things have you de-accessioned in the past 12 months?
   e. Do you consider PR issues when material is discarded? How do you protect yourself?

6. Costs.
   a. How much does it cost to maintain the collection and keep it accessible?
   b. What are the cost drivers in your environment?
   c. What criteria would you use to evaluate "scientific" or "commercial" value of data you hold or data that is offered to you?
   d. User fees: are they employed, or considered? If not, why not?

7. Managerial/Budgetary Issues.
   a. What are the main managerial/budgetary issues?
   b. How is success/failure defined with regards to managerial and budgetary issues?

8. Control of Holdings.
   a. Would you want to maintain control of your [things],

or would you be willing to have them managed by others (i.e. externally)? Why or why not?
   b. Please outline any collaborations that have worked well with other institutions/agencies?

9. Questions for State Geological Surveys only.
   a. How does your state benefit from your holdings? Can you give a specific example or anecdote?
   b. How would your state benefit from holdings you would like to have?
   c. Are there economic opportunities lost or public safety risks where data would have been useful?

Thank you.

D

Types of Geoscience Data and Collections

TABLE D-1 Examples of Geoscience Collections

Auger samples
Fluid samples (oil, gas, water)
Geochemical powder samples
Hand samples (incl. geotechnical, rock, and mineral)
Ice cores
Paleontological samples (micro/macro)
Rock cores
Rock cuttings
Sediment cores
Sidewall cores
Thin sections and polished sections
Type stratigraphic sections

TABLE D-2 Examples of Derived and Indirect Geoscience Data

Geophysical Data

Paleomagnetic resistivity
Potential fields
Seismic data (hardcopy, films, digital)
    Seismic refraction
    2-D and 3-D seismic reflection
Surface & airborne data

Velocity
Vertical seismic profiles
Well logs (paper, fiche, digital)

Petrophysical Data

Lithology logs (incl. mud and gas logs)
Routine (porosity, permeability, grain density)
Special (porosity & permeability under confining stress, Archie cementation, saturation exponent, capillary pressure, relative permeability)

Geochemical Data

Analyses—hard copy, digital

Other Data

Maps (topographic, geologic, subsurface, base, lease ownership, digital well spots, alteration, soils, groundwater studies, sample location, etc.)
Field notes
Paper reports
Photographs (aerial, satellite, slides, prints, planetary)
Scout tickets (fiche and paper)
Drilling/completion reports
Drill stem & other tests
Stratigraphic tops
Production history
Source-rock maturity analysis

E

Glossary

ACCESSION The process by which a specimen is formally entered into a collection. Accession includes listing the specimen in the collection's permanent inventory.

AQUIFER A body of sediment or rock that is sufficiently saturated to yield ground water to springs or wells.

BASIN A low area of the earth's crust in which sediments have accumulated.

BED Layer or *stratum*.

BIOSPHERE All the areas of the earth occupied by living organisms, including land, sea, and the atmosphere.

BLIND THRUST FAULT A thrust fault that does not reach the surface of the earth.

BOX (CORE) Storage containers for cores. A widely used size is approximately 3 feet long and holds three to five lengths of core (9 to 15 linear feet). Depending on the density of the rock, such a box can weigh 35 to 50 pounds.

BY-PASSED PAY ZONE A zone within the bedrock in which oil or gas resides, but which was overlooked in an initial analysis and extraction effort.

CATALOGING The process of recording metadata in some centralized database, usually with some kind of index numbering system, in any medium.

CEMENT Material that fills open space in sediments.

CLIMATE The characteristic weather over a region averaged over a long period.

COLLECTION A group of objects organized for ready access and study. Geoscience collections are groupings of individual geoscience items that may be related by sample type, geographic location, or scientific or applied interests. Museum collections commonly contain specimens of local interest or specimens that reflect the research interests of curators of the museum.

COMPLETION RECORDS/CARDS Descriptions of the engineering characteristics of a given well.

COMPUTERIZATION Informal term for the process of converting catalog data or metadata into a digital form.

CONNEX CONTAINERS Large transport containers used to store geoscience data and collections.

CORE A long cylindrical sample of the earth's crust (about 2 inches in diameter) taken most commonly by means of a diamond core drill (for rock) or by vibrating very rapidly a long metal tube (unconsolidated sediment cores).

CT SCAN Computed or computerized tomography; an imaging method that uses computers to assemble multiple x-ray images into a three-dimensional picture of the subject being scanned.

CURATION The process by which specimens are cleaned, sorted, boxed, identified, labeled, catalogued, and perhaps undergo reconstruction when they are properly brought into a collection.

CUTTINGS Cuttings are the chips of rock that come up the outside of the drill stem, after having been cut by a rotary drill bit. They are samples of the rock through which the drill bit has passed.

DATA Individual facts, figures, or other items of information organized for analysis.

DATA MIGRATION The transfer of data from one medium to another to ensure that they remain accessible as technology changes and older technology becomes obsolete.

DEACCESSION The procedure by which a specimen is formally and permanently removed from a collection.

DESICCATION The process of dehydration.

DISAGGREGATION The process by which particles break up or apart.

DISCOVERY The process by which the existence of data or information is gleaned.

DRILL WELL A hole created by the process of drilling into the ground

ENDANGERED COLLECTION A collection that lacks curatorial support at the moment and is in imminent danger of permanently losing curatorial support.

ENHANCED OIL RECOVERY (EOR) Techniques define subsurface fluid injection processes into hydrocarbon reservoirs to attain additional oil beyond that recovered by primary and secondary water and/or gas processes. The common methods are chemical flooding, gas (miscible) injection and thermal. EOR, also called tertiary recovery, supplement available reservoir energy, improved drive mechanism efficiency and/or production rate. Chemical flooding is an EOR method that involves the addition of chemicals to injected water or gas to recover additional oil by reducing the mobility of the injected fluid, or reducing the oil/water interfacial tension or both. Three common recovery processes are alkaline, polymer and surfactant/polymer flooding.

FGDC Federal Geographic Data Committee, established by the Office of Management and Budget to address the coordination and standards objectives in the development of the National Spatial Data Infrastructure (NRC, 1994).

FIELD A region or area that possesses or is characterized by a particular oil, gas, or mineral resource.

FORMATION Any igneous, sedimentary, or metamorphic rock represented as a unit or any sedimentary bed or consecutive series of beds sufficiently homogeneous or distinctive to be a unit.

FOSSILS The remains or traces of living things from the geological past preserved in the sedimentary rock. They include a huge variety of objects from dinosaur bones and footprints to petrified wood to impressions of shells on large rock slabs to the remains of single-cell organisms mounted on microscope slides.

GAS Hydrocarbons that exist as a vapor at ordinary temperatures and pressures.

GEOCHEMISTRY The study of the Earth through chemical methods.

GEOINFORMATICS The use of multi-disciplinary databases that facilitate the extraction of knowledge from the geologic record.

GEOPHYSICAL TRACKLINE A line along which a geophysical trace, or data position that can be read, occurs.

GEOPHYSICS The study of the Earth through quantitative, physical methods. Sub-disciplines include seismology and tectonophysics, among others.

GEOPRESSURE TOPS Tops of an anomalous subsurface pore pressure that is higher than the normal, predicted hydrostatic pressure for a given depth. Abnormally high pore pressure might occur in areas where burial of water-filled sediments by impermeable sediment such as clay was so rapid that fluids could not escape, and the pore pressure increased with deeper burial. Excess pressure, called overpressure or geopressure, can cause fluids to escape rapidly leading a well to blow out or become uncontrollable during drilling. SOURCE: Schlumberger Oilfield Glossary, 2002.

GEOSCIENCE(S) A short term for the collective subdisciplines of the geological (solid Earth) sciences, including engineering geology, geobiology, geochemistry, geohydrology, geophysics, sedimentology, and stratigraphy, among others.

GEOSPATIAL A term applied to geographic data that are spatial (they show the geographic distribution of a phenomenon).

GEOTECHNICAL The application of scientific methods and engineering principles to the acquisition, interpretation, and use of materials from the Earth's crust.

HYDROCARBON Any organic compound—gaseous, liquid, or solid—consisting solely of carbon and hydrogen. Often used as generic term for oil and gas.

HYDROCARBON CONTACT The point at which the oily hydrocarbon floats on the water of a reservoir. The formation still has irreducible water saturation. The gas-cap in the reservoir is also an example of a contact.

ICE CORE Long thin column of ice collected by drilling from a glacier or ice sheet, usually 2 to 6 inches in diameter.

INDEXING The process of creating a searchable tool for a collection, work, or document.

INDUSTRY The petroleum industry, including its various business segments such as exploration, production, refining, transportation, and marketing.

INSTITUTIONAL MEMORY The collective knowledge and history of an organization held by employees of that organization (institution), especially those who have been there for a number of years.

INTERSTITIAL Lying between particles comprising the matrix of a sample.

INVERTEBRATE An animal without a spinal column or backbone, includes organisms such as insects, shellfish and worms.

LITHOLOGY The character of a rock formation or a rock formation having a particular set of characteristics.

LOG See WELL LOG.

LOT A set of specimens collected in one place at one time.

MAGNETIC TAPES A tape that has a magnetic coating used for recording numeric data.

MARGINAL WELLS A low-producing oil or gas well which borders on being economically feasible to operate.

METADATA Term used to describe a dataset and bring value to the scientific data represented. Examples include collecting conditions, instrumentation parameters, location, depth, range, and the names of the analysts and techniques they employed.

MINERALOGICAL Pertaining to the scientific study of various inorganic natural substances and their properties.

ORPHANED COLLECTION A collection of scientific value that is no longer wanted by the institution or individual that houses it, and the institution or individual, either publicly or de facto, has renounced its responsibility to care for the collection.

OXIDATION The process of dehydrogenation especially by the action of oxygen.

PALEONTOLOGICAL Pertaining to the study of life in past geologic time, based on fossil plants and animals.

PERMEABILITY A measure of fluid flow or deliverability of fluids from a rock; it can only be measured directly from the examination of actual rock samples such as cores. Fluids contained in the pores of a rock cannot flow if it lacks sufficient permeability.

PETROLOGICAL Pertaining to the scientific study and classification of rocks.

PETROPHYSICAL Pertaining to the study of the physical and chemical properties of rocks, especially as it relates to their fluid holding properties.

POROSITY A measure of the fluid storage capacity of a rock; it can be determined by examination of cores or subsurface data.

PRESERVATION Various steps necessary to care for geoscience data and collections including: data acquisition, organization and maintenance, making users aware of samples and data, making data accessible, and assuring that data are useful and of sufficient quality.

PROFILE (PROFILING) The process by which a collection is evaluated against specific criteria. For example, a profile indicator or criterion might consist of conservation status, arrangement, or storage container quality.

PYROCLASTIC FLOW An avalanche of hot ash, pumice, rock fragments, and volcanic gas that spills down the side of a volcano at a rate of 100 km/hour or more. The temperature within these flows may be greater than 500° C. Once these flows have been deposited, they may flatten and weld together as a result of the intense heat and the weight of the overlying material.

REPOSITORY A storage facility that may or may not be climate controlled. For instance, a core repository will contain cores in addition to cuttings, logs, and other samples that either directly or indirectly augment the core collections themselves. Geoscience repositories contain both geoscience data and collections.

RESERVOIR A porous and permeable mass of rock that contains and/or transmits fluids.

SCOUT TICKETS A summary of a well's important information (e.g., drilling rates, total depth, production rates) prepared by a scout, an individual employed to gather such information, often from a competitor, by all available means.

SECONDARY RECOVERY Production of oil or gas as a result of artificially augmenting the reservoir energy (drive), as by injection of water or other fluid.

SEDIMENTARY ROCK Rock resulting from the consolidation of loose sediment that has accumulated in layers.

SEISMIC PROFILE A picture of the Earth's crust results of sending vibrations (produced by explosions or mechanical devices) into the Earth's crust. Different layers within the crust reflect these vibrations in different ways and thereby allow scientists to develop a picture of the structure of the crust across wide areas.

SERVER FARM A collection of servers that exchange large volumes of data and information across the Internet.

STRATIGRAPHY A sub-discipline of geology that deals with the origin, composition, distribution, and succession of strata or the overall arrangement of strata.

STRATUM (pl. strata) A layer of sedimentary rock visually distinguishable from other layers above and below.

SUPERGIANT A contiguous surface area beneath which one or more petroleum reservoirs either has produced or is expected to produce more than 5 billion barrels of oil or more than 30 trillion cubic feet of combustible gas.

SYSTEMATICS The study of the types and diversity of organisms and their relationships.

TERRAIN A tract or region of the Earth's surface considered as a physical feature or other distinguishing characteristic.

THIN SECTION A fragment of rock mechanically ground to a thickness of about a thousandth of an inch (0.03 mm) and mounted on a glass slide for microscopic examination.

THRUST FAULT A fault with a dip (incline) of 45 degrees or less on which the overlying side of the fault appears to have moved upward relative to the underlying side.

TRACKLINE The location or route followed by a seismic survey.

TYPE SPECIMEN The single specimen on which the original description of a particular species is based, which serves as a permanent point of nomenclatural reference for the application of the name of that species.

VUG A small unfilled cavity in rock.

WAREHOUSING The simplest form of storage; does not include curation of samples, or promoting accessibility.

WATERFLOODING A method of *secondary recovery* in which water is injected into an oil reservoir to force additional oil out of the reservoir rock and into the well bores of producing wells.

WELL A bore hole sunk into the ground for the purpose of obtaining fluids such as water or oil and gas.

WELL LOG A graphic paper or electronic record of remotely measured observations or tests made on the rocks through which the drill passed plotted as a continuous function of depth.

SOURCE: DOE, nd; Sheriff, 1994; and Jackson, 1997.

F

Acronyms and Abbreviations

AAPG	American Association of Petroleum Geologists	FLMNH	Florida Museum of Natural History
AGI	American Geological Institute	FMNH	Field Museum of Natural History
AGU	American Geophysical Union	FY	fiscal year
AMNH	American Museum of Natural History	GCR	Gulf Coast Repository
ANSP	Academy of Natural Sciences Philadelphia	GLF	Geophysical Log Facility
API	American Petroleum Institute	GMC	Alaska Geologic Materials Center
ARCO	Atlantic Richfield Company	GP	Geco-Prakla Inc.
BCR	Bremen Core Repository	GSA	Geological Society of America
BEG	Bureau of Economic Geology, University of Texas at Austin	HGRC	Herold Geological Research Centers (Denver, Houston, Abilene, Casper)
BELI	Balcones Energy Library Inc.	ICWG	Ice Core Working Group
BLM	Bureau of Land Management	IHS	Information Handling Service Energy Group (formerly Petroleum Information [PI], Dwights and Petroconsultants)
CAD	Canadian dollars		
CGSI	Cambe Geological Services Inc.	IMLS	Institute for Museum and Library Services
CIPA	California Independent Petroleum Association	IODP	Integrated Ocean Drilling Program
		IOGCC	Interstate Oil and Gas Compact Commission
CNWRA	Center for Nuclear Waste Regulatory Analyses	IOGS	Independent Oil and Gas Service (Kansas)
		IPAA	Independent Petroleum Association of America
CRC	Core Research Center (U.S. Geological Survey)	IRIS	Incorporated Research Institutes for Seismology
CSPG	Canadian Society of Petroleum Geologists		
DAAC	Distributed Active Archive Centers (NASA)	JLL	Jackson Log Library
DERL	Denver Earth Resource Library	JOI	Joint Oceanographic Institutions, Inc.
DOE	Department of Energy	JOIDES	Joint Oceanographic Institutions for Deep Earth Sampling
DOI	Department of the Interior		
DOSECC	Drilling, Observation, and Sampling of the Earth's Continental Crust	KE EMu	KE Software Electronic Museum Management System
DSDP	Deep Sea Drilling Project	KGS	Kansas Geological Survey
EAR	Division of Earth Sciences (NSF)	KU	University of Kansas
ECR	East Coast Repository	KUMIP	University of Kansas Museum of Invertebrate Paleontology
EIA	Energy Information Administration (DOE)		
EII	Energy Information Inc.	LACM	Los Angeles County Museum of Natural History
EOR	Enhanced Oil Recovery		
ER	engineering regulation	MCZ	Museum of Comparative Zoology
EUB	Energy and Utility Board of Alberta	MEL	Midland Energy Library
FGDC	Federal Geographic Data Committee	MGG	Marine Geology and Geophysics Division

MMS	Minerals Management Service	**ORNL**	Oak Ridge National Laboratory
MOU	memorandum of understanding	**OSM**	Office of Surface Mining
MSHA	Mine Safety and Health Administration	**PII**	Petroleum Information, Inc.
MSRP	Mineral Science Research Program (USGS)	**PRI**	Paleontological Research Institution
NARA	National Archives and Records Administration	**PTTC**	Petroleum Technology Transfer Council
NASA	National Aeronautics and Space Administration	**RELI**	Riley Electric Log Inc.
		SeaSat	sea satellite
NCDC	National Climatic Data Center	**SEG**	Society of Exploration Geophysicists
NESDIS	National Environmental Satellite, Data, and Information Service (NOAA)	**SELGEM**	Self Generating Master (Smithsonian database program)
NGDC	National Geophysical Data Center (NOAA)	**SSPLA**	Southern States Professional Log Association
NGDRS	National Geoscience Data Repository System	**SUI**	State University of Iowa
NICL	National Ice Core Laboratory	**THUMS**	Texaco, Humble (now Exxon), Unocal, Mobil, and Shell (consortium)
NIMA	National Imagery and Mapping Agency	**TMM**	Texas Memorial Museum
NMBGMR	New Mexico Bureau of Geology and Mineral Resources	**UCMP**	University of California Museum of Paleontology
NMIMT	New Mexico Institute of Mining and Technology	**UMMP**	University of Michigan Museum of Paleontology
NMMR	National Mine Map Repository	**URI**	University of Rhode Island
NMNH	National Museum of Natural History	**URL**	Uniform Resource Locator
NOAA	National Oceanic and Atmospheric Administration	**USACE**	U.S. Army Corps of Engineers
		USBM	U.S. Bureau of Mines
NODC	National Oceanographic Data Center (NOAA)	**USFS**	U.S. Forest Service
		USGS	U.S. Geological Survey
NPS	National Park Service	**USGS(D)**	U.S. Geological Survey Paleontological Collection, Denver
NRC	National Research Council		
NSF	National Science Foundation	**USNM**	U.S. National Museum
OCGSL	Oklahoma City Geological Society Library	**USNRC**	U.S. Nuclear Regulatory Commission
OCS	outer continental shelf	**UW**	University of Washington
ODP	Ocean Drilling Program	**VMNH**	Virginia Museum of Natural History
OILF	Oil Information Library (Ft. Worth, Texas)	**WCR**	West Coast Repository
OILW	Oil Information Library (Wichita Falls, Texas)	**WOGCC**	Wyoming Oil and Gas Conservation Commission
ONR	Office of Naval Research	**YPM**	Yale Peabody Museum of Natural History
OOIP	original oil in place		

G

NSF Division of Earth Sciences (EAR) Guidelines for Geoscience Data and Collections Preservation and Distribution

This statement provides guidelines from the Division of Earth Sciences (EAR), National Science Foundation, for the implementation of the Foundation's Data Sharing Policy. The overall purpose and fundamental objective of these policy statements is to ensure and facilitate full and open access to quality data for research and education in the Earth Sciences. These guidelines are considered to be a binding condition on all EAR-supported projects.

The Division of Earth Sciences conforms to the following statement on sharing of research results and data (NSB-88-215; PAM Manual #10, VII, G.2b):

SHARING OF FINDINGS, DATA, AND OTHER RESEARCH PRODUCTS

The National Science Foundation advocates and encourages open scientific communication. The NSF expects significant findings from research and educational activities it supports to be promptly submitted for publication, with authorship that accurately reflects the contributions of those involved. It expects investigators to share with other researchers, at no more than incremental cost and within a reasonable time, the data, samples, physical collections, and other supporting materials created or gathered in the course of the work. It also encourages awardees to share software and inventions or otherwise act to make the innovations they embody widely useful and usable.

NSF Program management will implement these policies, in ways appropriate to the field and circumstances, through the proposal review process; through award negotiations and conditions; and through appropriate support and incentives for data cleanup, documentation, dissemination, storage, and the like. Adjustments and, where essential, exceptions may be allowed to safeguard the rights of individuals and subjects, the validity of results, or the integrity of collections or to accommodate legitimate interests of investigators.

The Division of Earth Sciences is committed to the establishment, maintenance, validation, description, and distribution of high-quality, long-term datasets. Therefore:

1. Preservation of all data, samples, physical collections and other supporting materials needed for long-term earth science research and education is required of all EAR-supported researchers.
2. Data archives must include easily accessible information about the data holdings, including quality assessments, supporting ancillary information, and guidance and aids for locating and obtaining data.
3. It is the responsibility of researchers and organizations to make results, data, derived data products, and collections available to the research community in a timely manner and at a reasonable cost. In the interest of full and open access, data should be provided at the lowest possible cost to researchers and educators. This cost should, as a first principle, be no more than the marginal cost of filling a specific user request.
4. Data may be made available for secondary use through submission to a national data center, publication in a widely available scientific journal, book or website, through the institutional archives that are standard for a particular discipline (e.g. IRIS for seismological data, UNAVCO for GPS data), or through other EAR-specified repositories.
5. For those programs in which selected principle investigators have initial periods of exclusive data use, data should be made openly available as soon as possible, but no later than two (2) years after the data were collected. This period may be extended under exceptional circumstances, but only by agreement between the Principal Investigator and the National Science Foundation. For continuing observations or for long-term (multi-year) projects, data are to be made public annually.
6. Data inventories should be published or entered into a public database periodically and when there is a sig-

nificant change in type, location or frequency of such observations.
7. Principal Investigators working in coordinated programs may establish (in consultation with other funding agencies and NSF) more stringent data submission procedures.
8. Within the proposal review process, compliance with these data guidelines will be considered in the Program Officer's overall evaluation of a Principal Investigator's record of prior support. Exceptions to these data guidelines require agreement between the Principal Investigator and the NSF Program Officer.

SOURCE: NSF/EAR, 2002.

H

Web Sites

American Geological Institute
  GeoTrek
  http://www.agiweb.org/agi/NGDRS/ (click start GeoTrek)

  National Directory of Geoscience Repositories
  http://www.agiweb.org/agi/datadirectory/

  GeoRef
  http://www.georef.org

Australia Department of Mines and Energy
  National Geosciences Databases
  http://www.ga.gov.au/oracle/

  New South Wales Digital Imaging of Geological System (DIGS)
  http://www.minerals.nsw.gov.au/explore/digs.htm

  Northern Territory Geological Survey Mineral Databases
  http://www.dme.nt.gov.au/ntgs/geoscience_info/mineral_datab.html

  Queensland Drill Core Samples in Regional Repositories
  http://www.dme.qld.gov.au/gsd/edc_regional.htm

  Victoria Department of Natural Resources and Environment Databases and Indexes
  http://www.nre.vic.gov.au/web/root/domino/cm_da/nrencor.nsf/frameset/NRE+Corporate?OpenDocument&[http://www.nre.vic.gov.au/search.html]

  West Australian Mineral Exploration Index
  http://www.dme.wa.gov.au/wamex/

Butte, Montana, Chamber of Commerce
  http://fp1.in-tch.com/www.butteinfo.org/Attractions/history.htm

Canadian Provinces Data Catalogs
  Core available at the Mineral Core Research Facility of Alberta
  http://www.ags.gov.ab.ca/ext/cgi/code/core/core_list2.exe

  Manitoba Core Listing
  http://www.gov.mb.ca/itm/petroleum/core/core_100.pdf

  Nova Scotia Natural Resources, Minerals and Energy Digital Geoscience Data
  http://www.gov.ns.ca/natr/meb/pubs/PUBS3.HTM#databases

  Ontario Earth Resources and Land Information System (ERLIS)
  http://www.mndm.gov.on.ca/MNDM/MINES/ERLIS/erlis_db.htm

Catalogue of Meteorites
  downloadable list
  http://www.nhm.ac.uk/mineralogy/grady/catalogue.htm

Department of Energy
  Glossary of terms
  www.eren.doe.gov/consumerinfo/glossary.html

Kansas Geological Survey
  Core Library Samples Index
  http://magellan.kgs.ukans.edu/CoreLibrary/index.html

Library of Congress, American Memory Project
  Collection Finder: http://memory.loc.gov/ammem/collections/finder.html

National Geophysical Data Center
  Marine Geology Inventory
  http://www.ngdc.noaa.gov/mgg/geolin/geolinsearch.html

  Index to Marine and Lacustrine Geological Samples
  http://oas.ngdc.noaa.gov/mgg/plsql/curators_lakes.search_screen

  Grainsize Database
  http://www.ngdc.noaa.gov/mgg/geology/grainsizesearch.html

  Marine Minerals Bibliography
  http://zenith.ngdc.noaa.gov/mgg/geology/mmdb/mmbib.HTML

  Ocean Drilling Core Data
  http://www.ngdc.noaa.gov/mgg/geology/odp/odpintro.htm

  Deck 41 Surficial Seafloor Sediment Database
  http://oas.ngdc.noaa.gov/mgg/plsql/deck41.search_screen

  GEODAS: Marine Geophysics online system
  http://www.ngdc.noaa.gov/mgg/gdas/gd_cri.Html

National Ice Core Laboratory
  http://nicl.usgs.gov
  Master Inventory List (downloadable)
  http://nicl.usgs.gov/master.htm

APPENDIX H

Ocean Drilling Program
   Log Database
   http://www.ldeo.columbia.edu/BRG/ODP/DATABASE/

   Janus : ODP Database
   http://www-odp.tamu.edu/database/janusmodel.htm

Oceanic Abstracts
   http://www.csa.com/csa/factsheets/oceanics.html

Oil and Gas Revenues
   Alaska
   http://www.dog.dnr.state.ak.us/oil/programs/royalty/revenues.htm
   http://www.tax.state.ak.us/SourcesBook/2001FallSources/
   V.%20Oil%20Revenue.pdf

USGS Core Research Center
   Cores and cuttings database by township
   http://geology.cr.usgs.gov/crc/wellreport.htm

Wyoming Oil and Gas Conservation Commission
   http://wogcc.state.wy.us/ (select "cores")